# The Recollections of Eugene P. Wigner

## as told to Andrew Szanton

# The Recollections of Eugene P. Wigner

*as told to Andrew Szanton*

Plenum Press • New York and London

Library of Congress Cataloging-in-Publication Data

Wigner, Eugene Paul, 1902–
    The recollections of Eugene P. Wigner as told to Andrew Szanton /
  Andrew Szanton.
        p.    cm.
    Includes bibliographical references and index.
    ISBN 0-306-44326-0
    1. Wigner, Eugene Paul, 1902–    . 2. Physicists--United States-
  -Biography.    I. Szanton, Andrew.   II. Title.
  QC16.W55   1992
  530'.92--dc20
  [B]                                                          92-17040
                                                                   CIP

ISBN 0-306-44326-0

© 1992 Andrew Szanton
Plenum Press is a division of Plenum Publishing Corporation
233 Spring Street, New York, N.Y. 10013

Printed in the United States of America

# *Tilling the Soil*
## *A Collaborator's Introduction*

*I* first met Dr. Eugene Wigner while working on an oral history project at the Smithsonian Institution in Washington, D.C. Funded by the Sloan Foundation, the Smithsonian's Museum of American History was building a video archive of the personal recollections of those who had worked on the Manhattan Project during the Second World War, building the world's first atomic bomb.

I was introduced to Dr. Wigner at a Smithsonian panel in Oak Ridge, Tennessee. A few months later, I spent two days in Princeton, New Jersey, and interviewed him twice, at length.

Eugene Wigner's head is large, with gray hair on the sides and a few stray wisps on top, a prominent nose, large ears, and watery blue eyes that sparkle. He dressed with a casual formality. His voice was wonderfully rich, heavily accented, full of humor and regret.

Dr. Wigner was extremely polite. Inside his own office, he asked my permission before removing his suit jacket. His brief coughing fits, he assured me, were involuntary. "It is my fault," he said. "But not my *conscious* fault." But the politeness was not distancing. He offered his attention with a shy sincerity that I found affecting.

By the end of the second interview, I was struck by how much vivid material was in this life—the political and intellectual tensions of turn-of-the-century Budapest; the inspiration of Albert Einstein; the tense, foreboding work on the atomic bomb. I knew that no major book had ever been written about Dr. Wigner, and I urged him to write his memoirs.

To this, Dr. Wigner gave a very firm "no." For the scientist, he said, one's work is all that should remain. The American notion of celebrity he found distasteful. He was quite content with his niche in the history of physics, and wanted no part of a memoir.

Or so I thought until one day around June 1988, when he wound up a routine telephone call by asking politely when I planned to come up to Princeton and help him write his memoirs.

I was taken aback. But I managed to say that I would like to do that very much and would begin planning immediately. At the end of August 1988, I said a fond goodbye to the Smithsonian and moved up to Princeton.

I conducted over 30 interviews with Eugene Wigner that fall of 1988, most of them in his office, on tape. When we met at his home, he seemed to dislike the recording apparatus. I took notes.

Our very first interview was in his office. I asked Dr. Wigner what he wanted from his memoirs. He spoke imprecisely, but at length and with feeling. It was clear that he wanted the book to convey the love of science that he and his colleagues had felt. He expected the book to evoke the spirit of his principal mentors and colleagues and of their scientific work. Dr. Wigner was troubled that his memoirs would inevitably center on himself. He said repeatedly: "I do not want fame."

The transitory quality of fame was a favorite theme of Dr. Wigner. In his own case, he seemed pleased enough that he would soon be forgotten. But that many of his scientific peers would also be forgotten he clearly regarded as a great tragedy.

Underneath all of the modesty, the continental manners, the deep reservoirs of politeness, I sensed that Eugene Wigner felt that he had never been fully understood. His passion for science, his personal experience, and his thoughts on the political side of human nature had nowhere been put on the record.

❧     ❧     ❧

"Who is Eugene Wigner, and why should I read a book about him?" I have heard those implicit questions many times over the last few years. Here I will try briefly to answer them.

Eugene Wigner is one of a generation of physicists of the 1920s who remade the world of physics. It was a generation that extended from Berlin to London to Pisa to Zurich—and to Budapest, though not quite yet to New York or Chicago.

The first physicists in this generation—Werner Heisenberg, Erwin Schrödinger, and Paul Dirac to name three—created quantum mechanics, throwing open dozens of fundamental questions. The ones who came just after scrambled to answer them and to pose further, more complicated questions. Eugene Wigner was in the second group. He has posed and answered some of the fundamental questions of twentieth-century physics.

Dr. Wigner would be an important historical figure for his contribution to physics alone. And since he was too modest to describe them much in the heart of the book, I should summarize them here.

In the late 1920s, Dr. Wigner laid the foundation for the theory of symmetries in quantum mechanics.

In 1933, he was probably the first to find that the force between two nucleons—very weak ordinarily—becomes very much greater when the nucleons are brought quite close together—a million times greater than the electrical forces between the electrons in the outer part of atoms.

Dr. Wigner showed that the most essential properties of the nuclei follow from widely recognized symmetries of the laws of motion. And he proved that his laws of symmetry apply equally to both protons and neutrons.

He extended his research into atomic nuclei in the late 1930s. These researches have shaped more than nuclear physics; they help to explain much of elementary particle physics.

Wigner also developed an important general theory of nuclear reactions. He contributed greatly to the practical use of nuclear energy. Much of his work in solid-state physics is still central to that field.

In physics, Dr. Wigner has shown a rare array of talents: brilliance in theoretical matters; brilliance in the laboratory; and a superb grounding in engineering.

❧　　❧　　❧

Between 1939 and 1945, a part of this great generation of physicists remade the world again. This time it was a far larger world that they changed—one of nations, armies, and peoples. They did it first by understanding that an atomic bomb could be built; then by arguing that it must be built in the United States, immediately; and finally by learning how to build it, on the fly, and under terrible pressure. Eugene Wigner was a giant of this undertaking as well.

It turned out that Dr. Wigner also played the game of politics, and rather skillfully for a recent émigré and a professed political amateur. In 1939 and 1940, he played a crucial political role in agitating for a Manhattan Project, a federal response to the splitting of the atom.

During the war, Dr. Wigner directed theoretical studies at the Metallurgical Laboratory in Chicago, creating a working nuclear pile for the production of plutonium, which yields atomic energy when it is induced to fission.

Dr. Wigner was crucial to the Met Lab's success in understanding plutonium production and developing plans for a water-cooled atomic pile. He promoted construction of a large-scale atomic pile to produce plutonium and developed superb engineering designs used by the DuPont Company to construct an air-cooled atomic pile.

And in the years since the war, Eugene Wigner has been a national figure in promoting the adoption of a major civil defense program.

❃ ❃ ❃

The war ended almost 50 years ago. Eugene Wigner is now an old man, but still a Hungarian. I am an American two generations younger. We have spent many hours together without closing much of the gulf between us. More than once, he has looked at me at the start of an interview and asked with real curiosity: "How old are you?"

Eugene Wigner himself does not look 89. He is about 5 feet 7 inches tall and 150 pounds, slightly stooped with age. Like many very old men, he often seems boyish. He is young enough to climb stairs two at a time and to be chagrined that he cannot descend in the same way. He rises from a chair easily and often walks around a room to help himself recall.

Remembering is hard for him, and when we reach a hole

in his memory he is apt to say, "That is a scandal!" He holds himself to a high standard. My mistakes, though, he meets with grace, often saying with a smile, "Worse crimes have been committed."

I have been gratified to find that Dr. Wigner's recollections in my own interviews consistently square with the written record of his past thought. His accounts of the people and events of his life have remained internally consistent throughout the three years in which this book was formed.

He loves Hungary dearly. When I mention that I want to ask about Hungary, he says, "Good!" and claps his hands. He is very curious about other parts of the world—what young people are doing, thinking; how societies are changing, learning from their mistakes. Unlike many wise people, he is unafraid to marvel.

Eugene Wigner is so deeply polite that his old friend Edward Teller has charged him with creating an absurd myth of Hungarian modesty. I once asked him how he could be so modest. He answered that he was not modest at all, and apologized for not being nearly as modest as he should be.

"I am often disagreeable," he added. I was taken aback, never having seen him act at all disagreeable. "When," I asked, "are you disagreeable?" "I am being disagreeable at this moment," he said. "You accuse me of being modest, and I disagree with you."

Once I ended an interview by telling him that I respected him. He answered, "That is not a good idea." "Why not?" I asked. His answer was plaintive and amused: "What does it *do*?"

Dr. Wigner judges things by the extent to which they are "reasonable." Hitler was "unreasonable" long before he murdered a single Jew, unreasonable and therefore untrustworthy. Europe has not always been reasonable in settling its disputes. The United States he finds far more reasonable.

Princeton, New Jersey is a quite reasonable town. Princeton University was once unreasonable in firing him. But it was reasonable enough to rehire him, and so it would be unreasonable for him to bear it any grudge. A university town is a quite reasonable place for an 89-year-old man to spend the last years of his life.

Dr. Wigner's office is jammed with journals—mostly of nuclear physics, mostly unread, to his great dismay. He can't bear to throw them away. And books: scientific works in German and Hungarian as well as English; biographies of great men; a copy of *You're Only Old Once!* by Dr. Seuss.

❊     ❊     ❊

I have been the primary organizer of this book. I have listened to all of the tape-recorded hours, reread my notes, read many articles written by Dr. Wigner and various interviews conducted with him. From them I have shaped this book. I have not only condensed Dr. Wigner's words but arranged them into themes, chapters, and subchapters. I have even clarified his words by adding a few of my own.

At Dr. Wigner's encouragement, I have added basic historical research to his account, especially in its personal portraits. I have fixed in time such things as the exact year that Leo Szilard took his doctorate or Werner Heisenberg left Leipzig for Berlin. To Dr. Wigner this is detail, and though he wants it rendered accurately, he is half-pleased and half-chagrined to find that to himself, it no longer matters.

What I have taken care *never* to do is to add substance. Every single observation in this book has been made at some time by Eugene Wigner. Wherever meaning was vague, I have always asked him to clarify himself. This he has greatly enjoyed doing.

In organizing the book, I have leaned a good deal on

various of Dr. Wigner's published interviews. The chapter on civil defense, for example, was built around an article that Dr. Wigner published in *The Technology Review* of June 1964.

The book has borrowed from interviews that Dr. Wigner did with Thomas Kuhn, available from the American Philosophical Society; with *New Engineer* magazine of November 1971; with *Science* magazine of August 10, 1973; and with *The New Hungarian Quarterly* of Autumn 1973.

Dr. Wigner's summation of Leo Szilard's life was built around his tribute to Szilard, published in 1969, in the National Academy of Sciences' *Biographical Memoirs.* His thoughts about the lives of Johnny von Neumann and Paul Dirac both relied on tributes to those men that he had written earlier.

Finally, the letters exchanged between Dr. Wigner and Michael Polanyi over 30 years, and kept at the University of Chicago, have also been very helpful. A more detailed account of these literary debts comes in the bibliography, which covers all of the materials used in preparing the book.

❊     ❊     ❊

I have certainly marked the project by the questions that I have asked and not asked. But Dr. Wigner speaks freely. I feel sure that he has said nearly all that he wants to say; perhaps my questions have sometimes led him to say more. He and his family have reserved the right to remove what they find objectionable. To their credit, they have rarely done so. This book could not have been published without the generous help of Mrs. Eileen Hamilton Wigner.

While Eugene Wigner has cheerfully allowed me to examine his life, he has maintained a healthy skepticism about the value of the project. He knows that memory is fallible and he conceives of people and human society as fundamentally inscrutable. "Human emotions are not very well described, foreseen, or interpreted," he warned me in one of our first interviews.

I have arranged the book's preface to mirror the interview sessions. The preface begins with the same themes with which Dr. Wigner began almost every one of our interviews: his own unworthiness, the greatness of certain of his mentors and friends, and the arrogance implied by creating a memoir.

I am part Hungarian myself. The name Szanton comes from "szanto," and Dr. Wigner noted on the day we met that in Hungarian a szanto is "one who digs the earth." In the writing of this book, I have certainly dug the earth. But the story and the words are his. The soil is his.

In his scientific brilliance, his dislocation from his ancestral country, and his merging contributions to science and warfare, Eugene Wigner has been a kind of emblem of the twentieth century. Fascism, communism, capitalism—he has seen firsthand all three of this century's great, warring political ideologies. Beyond that, he is an observant man, a deeply decent one, and a strong witness. I believe now, even more than I did in 1988, that Eugene Wigner's story richly deserves telling.

*Andrew Szanton*

San Francisco

# $\mathcal{P}reface$

$\mathcal{I}$ have never wanted to be famous.
Sometimes, I hope that my life will soon be completely forgotten. You may say, "Oh, that could not happen to a Nobel prize–winning physicist." But many important scientists of 100 years back have already been forgotten. People get forgotten.

Now, to forget all people and all of human experience disturbs deep, instinctive human emotions. But history grows constantly and we cannot collect and recall it all. Most of the past disappears, and rightly so.

The earth has 4 billion people, and more every day. Each of these 4 billion lives must hold some interest, yet no one can hope to know 4 billion biographies. More and more biographies are written, most of them poorly. The collective biography of man swells distressingly and is ignored.

I read very few biographies, and I am not sorry. Clearly, I should know my own life story, and those of my sisters and wives. Perhaps we should all know 100 of the world's finest books and discuss them for pleasure. A few of these hundred books would certainly be biographies. Isaac Newton would have a biography; so would Albert Einstein.

Eugene Wigner does not belong anywhere near Isaac Newton and Albert Einstein. So I should be content with my brief profile in American Men of Science.

But I must admit that I am not. This kind of profile is too narrowly objective: "Hungarian physicist. Born 1902 . . ." It does not say that I loved a certain problem, that I saw subtlety and great beauty in it. It says only that I worked on it. Someone who never knew me could have written such a profile.

So I have decided to speak for myself, to broaden and deepen the record of my life, and to preserve my own memories of people and events that have changed physics, technology, even warfare.

Many of the men I want to describe have already been nearly forgotten. I trust that Newton and Einstein will never be forgotten. But what about those of only slightly lesser rank: Max von Laue, Walther Nernst, Michael Polanyi, and Wolfgang Pauli? How many of those names are widely known? How many will be known in 50 years time? We should know them all, and many more.

I am a physicist, not a historian, and a physicist does not always know dates and authorships as well as an historian. All recollection is flawed and the memories of honorable people often do not agree. I am now a very old man, and though my recollections are as truthful as I can manage, they may now be more flawed than most.

Well then, what advantage does an old physicist have over a professional historian? I think he has a very great one. The physicist may not know every date, but he recalls the spirit of times long past, a spirit that decisively drove him and his colleagues to rare and unexpected feats.

This spirit is not easy for historians to grasp or to recreate. The historian conducts interviews to learn what the old physicist knows in his bones. The historian collects photographs, but however suggestive photographs may be, they never give a full picture of the past, of its convictions, its personalities and the forces that moved them. So perhaps an old physicist should speak.

I cannot claim to be objective, but I have tried to be fair and polite. I hope my recollections will not offend anyone. I cannot see why people write books meant only to generate controversy. But history is important and should be discussed honestly. If this memoir criticizes some important people, it is because I feel we should not conceal the tiny weaknesses of very inspiring people.

Neither can I call my story comprehensive. A memoir selects certain events from a life's experience and omits the rest. This memoir omits most of the details of my personal life: just how I became fond of my wife or quarreled with my sisters. These are the things of diaries, a form that seems to me far inferior to the memoir. Diaries seem too often to only trace the patterns of the diarist's unhappiness.

Perhaps I should never have created this memoir. To speak at length about oneself often seems not only conceited but foolish. I have an idea that most people cannot keep in mind more than about 12 figures from history. They must save room in their consciousness for the celebrities of their age.

But I am a man with certain weaknesses. Indeed, I have more weaknesses than most men my age. Not only do I forget easily and sing poorly, I also have a weakness for reflecting on some of the signal events of my life. I want to leave some small record of those reflections.

Several years ago, I asked my collaborator, Andrew Szanton, to help me assemble this memoir. Together, we have thoroughly explored the major themes and events of my life.

And I must confess: I have deeply enjoyed the investigation.

## *Acknowledgments*

*T*he writing of this book has been a long and complex project, and this list of debts is not complete. I want to single out for special thanks six people:

Donald Stokes played a central role in insuring that the book would be published, read the draft manuscript with great care, and made superb comments.

My parents, Peter and Eleanor Szanton, gave three years of support and detailed advice. My father gave especially fine draft comments. My mother saw certain problems before I did, and suggested solutions to them.

My editor, Linda Regan, kept constant faith in the book and guided me through the publishing process.

Greatest thanks of all must go to Dr. and Mrs. Eugene Wigner. Mrs. Eileen Hamilton Wigner gave her hospitality and assistance. She helped me to understand her husband, and her ultimate belief in the book has been crucial to its publication.

Eugene Wigner commissioned this book in 1988, and offered me full use of his books, articles, and personal and official files. He sat for 30 formal interviews and as many informal ones. He patiently answered every conceivable question. In a few short years, Dr. Wigner has taught me a great deal about the meaning of kindness and dignity.

At Princeton, my thanks to the staff at Firestone Library and Fine Library; Bruce Finney, Pat Barwick, and the other staff in the physics department office; and to the office staff at the Woodrow Wilson School of Public and International Affairs.

Thanks to the American Institute of Physics (AIP), in New York, for use of the Wigner transcripts and for information about relevant photographs. Special thanks at AIP to Dr. Spencer Weart, Marjorie Graham, and Douglas Egan.

At the University of California, San Diego, thanks to Geoffrey Wexler, Catherine Georges and Richard Lindemann; at the University of Chicago, Richard L. Popp; at Argonne National Laboratory, Pat Canaday and Cindy Wilkinson. Thanks to Elizabeth Carroll-Horrocks of the American Philosophical Society.

Thanks to the University of Wisconsin's Archives; at the American Nuclear Society, to Teri Jarvie and Silvana Specha; at the Oak Ridge Public Library, to Joan Kearney. Thanks to the Richard A. Gleeson Library at the University of San Francisco; to the main library at the University of California, San Francisco; to the San Francisco Public Library; and to the Library of Congress, Washington, D.C.

Many thanks to a series of writers and scientists for their encouragement and suggestions.

Thanks to Edward Creutz for his generous support, for answering my questions about what Eugene Wigner was like in the 1930s, and for his careful draft comments.

Thanks to William Lanouette particularly for professional advice and for valuable leads on the Eugene Wigner literature.

Thanks to John McPhee for many things: for the example of his own work, for the quality of his teaching about writing in general, and for encouraging this project in particular.

At the Oral History Program in the Smithsonian Museum of American History, thanks to Stanley Goldberg, James Hyder, Terri Schorzman, and Philip Seitz.

Thanks to other audiences and readers: Carol Cannon and her students; Randy Hostetler and Francesca Talenti; Trish Perlmutter, Tom Perlmutter, Carol O'Neill, and Mike Chase; Joseph A. Trunk; Ken Wong; and Brad Woods.

Thanks to Levente Csaplár for his fluent translations of Dr. Wigner's Hungarian letters; to Marc Pachter for his seminar on biography, to Yuval Taylor for years of precise encouragement, and to Katherine Wolff for her translation of several of Dr. Wigner's German letters.

Thanks to Kenneth Schubach, Susanne Van Duyne, and others at Plenum Press.

For general assistance, thanks to Hubert Alyea, Naomi Brier, Andras Csaplár, William Golden, Judy Nelson, John O'Donnell, Rick Ross, Judith Shoolery, and Prudence Steiner.

Thanks to others in the Wigner family including David Hamilton, Charles Upton, and Martha Wigner.

Deep thanks to others in my own family: Sybil Stokes, Nathan Szanton, Sarah G. Szanton, and Sarah L. Szanton.

Finally, thanks to my wife, Barbara Cannon, for years of support and assistance.

# Contents

# One

# *Pit-a-Pat, Pit-a-Pat*

*J* stood watching Enrico Fermi, in a large room beneath Stagg Field at the University of Chicago. It was midafternoon on Wednesday, December 2, 1942. The United States was fighting the Second World War. To win that war, we felt we needed to make an atomic bomb. And to make that bomb, Enrico Fermi was trying to initiate the world's first controlled nuclear chain reaction.

It was part of what we called the Manhattan Project.

About 50 people stood in a room 30 by 60 feet. In the middle of the room a huge pile of black graphite bricks and wood pieces had been heaped up. The bottom half of the pile was square shaped; the top was narrower. A gray balloon cloth covered this brick pile on three sides in case Fermi wanted to catch lost neutrons. Today, we would call this pile a nuclear reactor. In 1942, and for some years after, we called it simply a nuclear pile.

The 50 observers had been collecting since 8:30 that morning at one end of the squash court, on a balcony about 10 feet above the floor. We had watched Fermi work on that nuclear pile. The serious work had begun at about 9:45. The chain reaction had almost been achieved around 11:30, but a safety rod set too low had stopped it. Fermi had sent everyone to lunch, and we had all returned to the squash court at 2:00.

For the last few days, Fermi had been making a series of complex calculations about the pile. He had also helped carry and assemble its graphite blocks and columns. The design of the nuclear pile appeared simple, but it was not. Fermi had given the pile large cadmium-plated control rods to stop the chain reaction. The control rods were electrically operated and could be withdrawn by throwing a switch. Lights indicated their position.

Now Fermi stood with a slide rule at one end of the balcony, with two of his top assistants, Herbert Anderson and Walter Zinn. Next to them was Arthur Compton, the director of all Manhattan Project work in Chicago. All four men were checking instruments in the control cabinet.

The other 45 of us were crowded on the far side of the balcony. Waiting near me was Crawford Greenewalt, a top official at the DuPont Corporation. I had invited Greenewalt to prove to him that a nuclear chain reaction was possible. Among the others present was my old friend, Leo Szilard.

Enrico Fermi was one of the greatest men I have ever known, and deserves a fuller description. He was Italian, just past 40 years old on that day in 1942; about 5 feet 6, with a friendly, handsome face, bluish eyes, and balding, dark hair.

Eight years before, Fermi's theory of beta decay had greatly inspired me. In the mid-1930s, most of the progress in physics seemed to come from either James Chadwick's lab in

Cambridge, England or from Fermi's in Rome. Fermi had won the 1938 Nobel prize for work with neutron absorption and radioactivity.

But for all of my regard for his work in physics, I had never seen Fermi until about 1938, when he and his Jewish wife, Laura, had escaped the fascism of Italy. Though Fermi was not yet an American citizen, he was now happily placed at Columbia University in New York City and already felt himself an American. He was about to give something of signal importance to his adopted country.

Fermi had begun as a theorist, but by now worked almost entirely in the laboratory. I thought him the single most important member of the Metallurgical Lab: careful, imaginative, uncommonly clever, with a gift for improvising with the materials at hand. He had the theoretical skill to know what he wanted and the experimental skill to find it.

It was James Chadwick who had first discovered the neutron in 1932. But Fermi's team had found something further: that the neutron could be slowed down to make an atom artificially disintegrate.

In 1942, the nuclear chain reaction was assumed to exist. It was Fermi's job to find it. For at least a year, I had felt that the chain reaction would be established and that Fermi would lead the group to do it.

Now, what was I doing in this room? I was another member of the Metallurgical Laboratory, and I was leading a group designing another reactor, this one in Hanford, Washington. My job may have been more technically difficult than Fermi's because our reactor had to be very large and had to operate at a very high intensity for a very long time. That was quite an engrossing problem.

But we all knew the central importance of Fermi's work.

We tried to help him, but Fermi led with vigor and skill, needing little help. He was not a man to hesitate.

In the small international world of physics, Fermi's name commanded the greatest respect. So if he wanted a great deal of graphite, people were quick to get it to him. He did not have to give orders to his assistants. They soaked up his suggestions.

Fermi's most striking trait showed in both his scientific work and his human relations. It was a simple realism, a perfect willingness to accept facts and men as they were. Though he understood the most complex physical theories, he avoided using them unless they were clearly needed. Likewise, he did not spend a lot of time analyzing his colleagues and friends, but he very seldom misjudged them.

Fermi was not only able but affable, and, despite his fame, completely unpretentious. He had a temper, but he was careful to control it. I doubt that any of his close assistants ever disliked him. The weight of the Manhattan Project and the subtle competition between my group and Fermi's had not hurt our friendship. I could still approach him with a problem and still enjoy his sense of humor.

But during the project we had seen very little of each other. Fermi and his people had been so busy calculating and building the chain reaction. We had been very busy designing our own reactor. And Fermi had changed a bit; he was not quite his ordinary self. His humor was harder to see.

Needing a lattice framework for his pile, Fermi had constructed one empirically, testing various designs before building the one that promised to induce a nuclear reaction with the smallest amount of material. The question required a certain clever reflection, but it was hardly theoretical. Fermi chose a large lattice with a diameter of several meters.

At 3:30 PM, Fermi calculated the rate of the rising neutron count and instructed an assistant to pull out the cadmium-plated rods about 25 centimeters at a time. I expect he felt nervous but he did not show it. He had given the pile extra control rods in case of an emergency, and even positioned a few men above the pile with a cadmium–salt solution in case all the control rods failed to stop the reaction. Fermi was every bit in control, as he was of every one of his experiments.

The recorder started to tick: pit-a-pat, pit-a-pat, pit-a-pat. As neutrons were absorbed, the chain reaction was more and more closely approximated. The uranium collided with the neutrons, which made more neutrons.

For a time it died down, but when the whole control rod had been pulled out, the recorder ticked more strongly than ever. We knew that the nuclear reaction was achieved. We had released the energy of an atomic nucleus and had successfully controlled that energy. Fermi liked to compare the chain reaction to the burning of a rubbish pile: small parts igniting other small parts, until the pile exploded in flame.

There were smiles all around the room, even some applause. But mostly we watched. For about 30 minutes, we watched. The world's first nuclear chain reactor was in operation. It was not theatrical or striking. Fermi had made the reactor quite weak so that it would not kill us all. But it was there; and it worked. Just before 4:00 PM, Fermi ordered the reaction stopped. The control rod was replaced and the reaction halted.

Enrico Fermi was not the only man who could have created this chain reaction, but perhaps the only one who could have done it so quickly. We had all expected the experiment to succeed. After all, if a carriage is built and hitched to a team of horses, we expect the carriage to move. Fermi had built the carriage and hitched the horses.

But as it moved, we were spurred on not only by the

strange excitement of nearing a working atomic bomb, but also by that familiar pleasure that comes with the success of any new scientific idea. Just over ten years since Chadwick had proved the existence of the neutron, we had achieved a controlled nuclear chain reaction. We knew that we had really changed physics.

Expecting this moment, I had bought a bottle of the Italian wine, Chianti, about ten months before in Princeton, New Jersey, and brought it to Chicago. I had guessed that the war might prevent Italy from exporting Chianti. In a way, it was harder to foresee that shortage of Chianti than to foresee the successful chain reaction. But I had been through the First World War, and I knew that such luxuries disappear.

Chianti was a special wine to me. I had first tasted it one summer in Venice around 1913, when I was 10 years old, a child traveling with my parents. The memory of those first few cubic centimeters of wine had remained vivid in my mind.

I had kept the Chianti behind my back throughout the experiment. Now I produced it from a brown paper bag on the balcony floor, brought it forward, and presented it to Fermi. He thanked me. He seemed to enjoy the exchange and to appreciate the preparation and goodwill involved, though what went on inside his mind, I could not know.

Certainly, Fermi knew that this moment had been built on the achievements of many other scientists: Henri Becquerel, who had discovered radioactive elements in 1896; Pierre and Marie Curie, who had found radium in 1898; Albert Einstein, who had told us in 1905 that mass was equal to energy, suggesting that one could be made into the other; Ernest Rutherford, who had discovered the nucleus at the atom's core in 1912; James Chadwick, who had found the neutron in 1932; and Otto Hahn and Fritz Strassmann, who had found out how the uranium atom can be split. Perhaps Fermi thought of these great scientists; he was careful to credit each of them when he spoke about the successful chain reaction in later years.

We all stood around, reassured, pondering the implications of success. There was no more chance of a conceptual error. We had moved matters to the next stage. For the first time, we could expect ample financial support for the making of the atomic bomb.

Fermi uncorked the bottle and asked someone to find paper cups. They were produced and we drank the sweet red Chianti. What a beautiful, subtle pleasure wine gives! Fermi signed his name just below the top of the Chianti label. The bottle went around the room, and below Fermi's signature the rest of us added our names.

The bottle ended up in the hands of Albert Wattenberg, a bright young physicist. No written records were made of the witnesses to this historic moment, so the names on the Chianti label were later used to recreate the group.

Quietly, we toasted the event, and wished that somehow this nuclear reaction could make man's life happier, and humankind less prejudiced.

But the moment was not completely joyous. There was also a pressure there, a weight. A chain reaction itself is a somewhat frightening thing. As it multiplies exponentially, one feels the tiny fear that it will overwhelm its controls. The thought of repeated chain reactions suggests that someday a nuclear reaction could overwhelm its human creators.

Every man in that room knew people who did not approve of what we were doing. The great German physicist James Franck dearly hoped that nuclear bombs would prove impossible to build. My own wife, who was also a physicist, felt the same way. So did Einstein. We all felt, in some way, that it would be a better world if nuclear energy were not so easily created.

So, as we drank the Chianti, I could feel silent prayers going up, prayers that building the atomic bomb was the right thing. We knew that man's inventions can have bad consequences as well as good. We knew that an atomic bomb could kill millions of innocent people.

I doubt that a single person in the room voiced his doubts about the atomic bomb. But, you see, none of us had to. We knew each other well and could sense them.

The physics of a nuclear chain reaction was already known. But the psychology of making an atomic bomb had yet to be explored. So the creation of the first controlled nuclear chain reaction was quite interesting from a psychological aspect as well as a military one.

Making a great weapon is not something to be proud of.

# Two

# Be a Good Son. Obey Your Mother Carefully . . .

*I* am a very old man now; my life began in Budapest, Hungary, on November 17, 1902. It was a completely different world then. Middle-class families lived quite happily without automobiles or radio, gas or electricity.

In the year 1902, humans had lived on our earth for millions of years, but science was very new. Modern physics had only begun about 200 years before, with the work of Sir Isaac Newton. And modern chemistry was only about half as old as that; John Dalton had begun much of it about 100 years before.

The great scientists of 1902 were largely content without atomic theory, quantum theory, or relativity theory; without any real knowledge of the nucleus—with just an inkling of modern physics. Their minds roamed a scientific world entirely different from the one we know today. And yet there were important scientists in 1902 who felt that everything important in science had already been discovered.

My parents named me Wigner, Jenö Pal, pronounced "*Vig*-ner *Yen*-nuh Paul." I was often called Jancsi ("*Yahn*-shee"). I am "Eugene Paul Wigner" today only because Wigner, Jenö Pal is not a good American name. Actually, it was hardly a good Hungarian name. "Wigner" is German, and "Jenö" is a rare name even in Hungary. If it had a name day, it was barely celebrated.

What a pity it is that we cannot recall the day of our birth. What a memory that would be! Like all children, I was born without my permission. But as soon as I realized that I was alive, I was curious about the world and happy to inhabit it. At least internally, I thanked my parents for having given me life.

My first clear and important memory is of the day that I discovered speech. I was about 3 years old. I had been genetically endowed with some interest in the sounds coming from the mouths of my parents. I had heard Hungarian spoken ever since I first drank milk from my mother's breast. But until that day in the country, all those Hungarian words had left a scant impression. I had never wished to imitate the sounds I heard.

On that magical day, I was walking in the southern Hungarian country near a place called Belcza-Puszta. I had come with my mother and father on a visit to the great estate of my uncle Kremzir, near where a large river meets the Danube.

Uncle Kremzir's estate had vast wheat and corn fields, a beautiful garden, horses and other animals, a small factory, and even a rude automobile garage.

The day was pleasantly warm. We walked together on a path beyond the wheat fields. I heard the familiar hum of voices, and abruptly I realized that I liked to talk, and even more to hear the jokes of my uncle and grandfather. Those two men were telling a series of jokes, each with a pleasantly abrupt conclusion, showing man's weakness, sometimes their own. We all laughed heartily. That simple memory has remained with me for 85 years.

Now, why do humans love jokes, which serve no practical purpose? Food and shelter are necessities. But laughter is not. So why do we invent jokes with such skill, and laugh at them with such pleasure?

Jokes are apparently universal, but no country could possibly love them more than Hungarians did at the turn of the century. I have never known such a taste for jokes in all the years since I left Hungary; certainly not in Germany and not in the United States either. But then perhaps we recall the things of childhood with a special fondness.

❊     ❊     ❊

My parents were Jewish, well-to-do and reasonable people, sober of alcohol, happy, and living together. My childhood world was devoted to reason. My father had a stable administrative job in an important leather tannery. My parents treated us reasonably and rarely punished us. We tried to help them. If a child wanted something, he had to ask permission. And if his request was reasonable, permission was granted.

Our schools taught us what was reasonable; our friends and neighbors acted reasonably toward us and expected the same in return; the great majority of people I knew were friendly and intelligent, without hatred or power lust. The whole flavor of life in Budapest was then thoroughly reasonable.

I greatly admired my father and deeply loved my mother. My two sisters I loved intensely. My older sister Bertha, called Biri, was a darling, undemanding child, with hair far down her back. Biri and I were especially close. Until she was nearly 20, she was a superb student at a local girls school. Then she left home and school to marry.

Margit, called Manci, was two years behind me. Manci produced less hair than Biri, though hers too went past her shoulder. Manci hardly obeyed the rules of the game. She was natural and impulsive, and even as a young girl she would not be ruled. Her marks in school were only average. And though she never openly defied my father, she sometimes ignored his wishes. Manci did not always ask permission.

As little children, Manci and I often quarreled. She seized my handkerchief, or I took some sweet of hers. A swift accusation and a quarrel ensued. But after a few years, this unhappy pattern ceased and we became dearly fond of each other.

My parents provided for us generously. I wore short trousers and a linen coat in summer; long stockings and a leather coat in winter. Not an exciting wardrobe, but then the clothes of children were not meant to attract attention.

We ate well: a plain breakfast before school and work, then at ten o'clock a larger meal, perhaps bread with ham. School ended at midday, and we came home for our principal meal: often soup, then beef, chicken, fish, or goose, with spinach or potatoes.

We had a well-laid table with fine tablecloths and napkins, china, and silverware. We ate sweets for dessert, often honey pastry. We ate another meal in the late afternoon and a fifth and last meal at 6:30. Our governess joined us at the table; the maids did not.

We hardly spoke at mealtime. As we sat down, my parents appealed to the Lord: "Help us Lord, and we admire you." But after that, religion was quite absent.

We might have discussed politics. The Parliament building was near our home. But we never made any effort to meet the politicians or even to read their speeches.

Politics was not a Jewish business. About 800,000 people lived in Budapest then, about 200,000 of them Jews. But there was just one Jew in the Hungarian Parliament, and his presence in that body was considered a miracle.

My parents did not discuss the absence of Jews in national politics, just as they did not discuss whether a sexual interaction is needed to produce children. Such topics need not be expressly forbidden; they simply do not arise. Children sense quickly what is taboo.

Meals at home were not meant for conducting advanced discussion; they were meant for proper eating. We were grateful to eat; food was expensive and poverty quite common and serious. So we sat at the table and fed ourselves industriously.

If my father asked for the day's happenings in school, I told him, though I never had a quick tongue and sometimes expressed myself awkwardly. My father spoke little of his own work, and my mother and sisters were quiet. But it was homelike. After dinner, my father might smoke a Hungarian cigarette or play cards. A few members of the family would go for a walk, and when we returned, the children were sent off to bed.

We had electricity at home. We had music, too, but no great devotion to it. We played Beethoven on the phonograph, but phonograph records were scarce. My parents preferred simple Hungarian folk music. But my father never sang and my mother quite rarely. My own singing was miserable. I knew that singing was important and had heard other people sing octave intervals. But I could never find the right sound, and after a time I did not dare try.

My mother played the piano, and a woman came to the house and taught Biri and me to play the piano, too. A few times a year, we went to the theater as a family, but almost never to the cinema.

Our houses were well furnished and often smelled of

roses. French relatives of my mother sent us flowers. I shared a bedroom and tiny playroom with my two sisters. My favorite room was the library, with its poetry books. I knew that having a private library was a treat, and I often came in to borrow books.

It is possible for a man of little education to be cultured, and that is what my father was. He barely had time to read, but he subscribed to three newspapers: a daily, a weekly, and a monthly.

Around 1914, my father bought a complete set of the novels of Mór Jókai. Jókai had died just ten years before, and his stories, poems, and dramas were revered as Hungarian national treasures. His romantic novels were his best. He was, so to say, King of the Hungarian novel.

Jókai was enormously prolific, with about 110 full-length novels. I had no hope of reading all of his works, but I enjoyed a few of his classics. They evoked Hungarian life with a delightful impressionistic style. Most of them described a world far better than our own.

Education was important at home. My first school grade was the third grade. Before that, my mother hired a woman named Gita to teach us at home. I began lessons at about age 5. Gita taught reading, writing, and arithmetic quite well. She rarely gave us tests, but she knew how much we knew.

Gita praised me, but modestly. She never predicted that I would win any prize. Such things were not done, so that small boys would not become conceited or self-conscious. Likewise, our plainness or beauty was never mentioned. So as children we had very little idea of how we might behave as adults. Our job was to learn and to remember, not to gaze into the distant future.

I could read by age 6, but I rarely did. The books my father pushed at me seemed lifeless. I was just a young boy with hair falling down my forehead and an interest confined to the moment.

I played several different Hungarian games similar to chess. I liked running but I was never swift. My eyeglasses made me an awkward sportsman. And I was a small boy. For exercise, my family took long walks and swam in the Danube north of Budapest.

When I was about 9, Gita received a high school teaching position, and left our family. I was not truly sad; 9-year-old boys do not grieve at the loss of a teacher. They know that many other teachers will follow. But in some way, I missed Gita.

In grade school, we learned the small multiplication table: $1 \times 1 = 1, 2 \times 2 = 4, 3 \times 3 = 9$ . . . all the way up to 9. The whole class used to shout it in unison.

❊   ❊   ❊

Now, let me forget my family for a minute and tell you a bit of Hungarian history. Hungary's first inhabitants were related to the Romans. But an alliance of Magyar tribes came from East Asia around 896 and occupied Hungary. That no doubt angered its inhabitants. But I should explain why the Magyars came.

Invading armies have always existed. But two basic types of invasions should be distinguished. In the first kind, the invader's homeland is too small, too short of food and other necessities. He invades in order to secure them. The second kind of invader thinks chiefly of ruling a great empire. He wants domination, pleasure, and glory. He pillages needlessly. The invasions of the Persians against the Greeks were of this kind, as were Hannibal's campaigns against the Romans. Adolf Hitler was this kind of invader and so was Joseph Stalin.

All of them were inspired chiefly by the dream of a great empire.

The Hungarian invasion of 896 was surely the better kind. The East Asians needed land and food. And they conquered fairly peacefully. They even brought gifts to the Hungarian ruler. He accepted them and thanked his visitors. And they said: "Well, now we have bought your country. We are the rulers here." And so they were.

The East Asians changed the national borders and brought new farming methods. Hungarian culture developed differently from German, French, or Italian. The Hungarian language is Finno–Ugric, lively and quite unlike the Romance languages, though apparently Finnish somewhat resembles it.

After a battle in 1526, the Turks occupied for a time the best regions of Hungary. And even after we had shaken off the Turks, Hungary was not an independent country, but a piece of the Austro-Hungarian empire.

In 1848, trying to break free from the Austrians, Hungary exploited political confusion in Vienna to make a bold series of national reforms. But a Croatian military leader invaded Hungary on behalf of the Austrians, and when he could not silence the independence movement, the Russians invaded as well. The Austrians made savage reprisals after the rebellion was quelled, and Vienna's grip tightened over Hungary.

After 1848, political and military power were hardly available to Hungarians. We turned our minds to better things. We studied how military power is consolidated and used. We celebrated our culture with folk songs and stories. We wrote superb novels and even greater poetry, perhaps the finest poetry in Europe.

Budapest grew to be our greatest city, a gorgeous city divided by the Danube River into two sections. West of the

Danube was the Buda section, east was Pest, and the two were connected by five famous bridges. Buda was the King's place, an entire district of castles. The handsome royal brick palace was on a mountain in Buda. It had just been restored in 1904. Buda had little ponds high in the hills. It was a showplace of great natural beauty. Pest was flatter, more common.

My family lived in Buda, first in a rented apartment in a pretty southern region; then in a plainer region closer to my school; and finally on the second floor of a beautiful home on the far side of the Danube, just off Andrássy Street. The house was joined to its neighbors, as many homes were in those days.

The older streets in Budapest were brick, the newer ones asphalt. Most of the prominent buildings were brick; wood burned too easily. People walked on the right side of the road. Horse-drawn buses and carriages used the middle. Automobiles were just beginning to be seen.

What impressive buildings we had! The Museum of Arts in Pest was filled with striking paintings. The grand Parliament building had just been finished. Inside fine two- and three-story exchange buildings, stock was bought and sold. In handsome churches, preachers gave their sermons.

Budapest had built a subway, the Millennium Underground, in 1896. It was the first in Europe, though the line was shorter then. You could still bargain in bookstores and groceries then. Many Hungarians spoke German; English was not yet popular. There were many cafes, of a kind that hardly exist in the United States. In such places, you were not only allowed to linger over coffee, you were supposed to linger, making intelligent conversation about science, art, and literature.

We celebrated Hungary's founding and other national festivities. The anniversary of our defeat in 1848 was a sad occasion. But our anthem was fine: "Dear Lord: Bless the

Hungarian with good spirit and much wealth. Extend a protection to him when he fights an enemy." It meant something to be a Hungarian.

    ✤     ✤     ✤

When my father wanted to take the family on a trip, he summoned the carriage driver, who brought the horse and carriage from a nearby stable. I knew this carriage driver a bit. When the carriage was crowded, I sat right next to him. But we had very little in common. I would have preferred talking with the horse, but he did not speak Hungarian.

One birthday, my parents gave me a first-rate bicycle: German or Austrian, all black but for the wheels, with both hand and foot brakes. I hardly needed a bicycle, but I liked to ride a few kilometers out from the city.

In summer and over long holidays, we often visited my mother's father, Doctor Einhorn. He lived in Kismárton, a town 200 kilometers west of Budapest. That does not seem a great distance today. It did then, when travel was far more risky and slow.

Kismárton was a pleasantly small town of about 30,000 people. The Austro-Hungarian government did not accord it much standing because culturally it was unmistakably German. Kismárton is now the east Austrian town of Eisenstadt, and its importance has been duly recognized.

Grandfather Einhorn was descended from Austrians living in Hungary. He observed Jewish custom a bit more than we did, but less than most Jews in Kismárton. He spoke far better German than Hungarian. My grandmother was his chief emotional support. She was a fine wife, but so retiring in manner that today I cannot even recall her face. They also kept a few servants.

My grandfather had a physician's general practice. He loved medicine. It exhausted him to practice as carefully as he did, especially in winter. People were always dying. So many children died that the average life span was under 30. To be a small-town doctor in that time and place was an act of great devotion, almost heroism.

Sometimes I wondered why my grandfather accepted that burden. But then he was deeply attached to Kismárton. He had grown up near there, and his practice taught him to know Kismárton far more intimately than the Wigners ever knew Budapest.

My grandfather spent mornings in a Kismárton hospital; afternoons, he returned to his gas-lit home. One room was his doctor's office; another his waiting room. When patients came in, he listened to their heartbeats and investigated their cares. He read the current literature on medication. A driver took him on house calls. He did light surgery, even a few abortions to save women's lives. But major surgery was quite another art, and one he avoided except in grave emergencies.

I enjoyed Kismárton. It was odd to hear more German spoken than Hungarian, but we managed. My mother knew German already. We grandchildren learned it to speak to our grandparents. German is a fine language. Foreigners often object to the length of some of its words, but to a native speaker, they do not seem overlong.

Sometimes, we traversed the small Jewish section of Kismárton. It must have been full of circumcisers, Yiddish newspapers, and kosher shops. But we hardly noticed these staples of Jewish life, and never engaged the Jews we saw on the street.

I was a born walker. By the time I was 10 years old, I was taking long, private walks. Every child experiences indignities of some kind; I found a solitary walk could ease the most intimate embarrassment. There were no questions to answer.

I also liked to walk with my family. And there were wonderful walks in Kismárton. To the right of the Einhorn house was a beautiful spring from which we gathered water in a stout bottle. Like most homes in Kismárton, the Einhorn house had no running water.

A walk to the left was even better, for soon we reached a museum and the grand palace of the Duke Eszterházy. Even a duke needs some privacy, and parts of Duke Eszterházy's palace grounds were closed to the public. But the Duke was a generous man, who liked to open his palace to public view. And much of his land was kept as an enormous public garden. We walked through this rich garden, sat on small benches, and chattered.

❖     ❖     ❖

Hungary was then changing from a feudal land to an urban one. But it was still largely agricultural. It had much coal and salt, a meager oil reserve, but a great deal of land. Two of my mother's sisters had married wealthy farmers. One uncle lived near Györ in western Hungary and rented a plot of about 500 acres. The other uncle was Uncle Kremzir of Belcza-Puszta, the one we were visiting when I first realized the full meaning of human speech. Uncle Kremzir owned about 1000 acres. One of my mother's sisters had become his second wife. He had about six children from his two marriages. The Kremzirs lived in a grand house on uncommonly elaborate grounds.

Uncle Kremzir had about 25 peasant families working

his land. They lived on the land, in a small village of huts. They knew the soil and all that farming required.

Uncle Kremzir had to know the land, too; to know when to plant corn, when to plant wheat, rye, oats, and barley. Perhaps the new field should be planted with potatoes or tomatoes. He rotated his crops so that the fields never had the same crop two years running. And he also had to direct his peasants.

My uncle owned a refinery on his land. He bought petroleum privately, separated it into oil and gasoline, and sold both parts. These and other business affairs kept him busy and a bit apart from his children and his nephew, Jenö Wigner. But I felt that I knew him and liked him dearly.

I liked watching his peasants, too. They were good Hungarians, who loved speaking Hungarian and being Hungarian. They were patriotic enough to risk their lives in a long, violent war for Austria–Hungary. But they had no desire to stretch the borders of their country. They fought when they were told to fight and they fought well. But they knew that war is a horror.

I dug up some earth next to these peasants, even planted a few things. But not seriously. That was the peasant's job, and we both knew it. These people were quite friendly, and if they resented the wealth of my uncle, they were careful not to show it. My uncle was a "progressive" sort of landowner, the kind of man who enjoys mingling with his peasants and asking their views on various farming methods.

But the peasants rarely spoke to my uncle or his family, unless they had been spoken to. They had worked this land nearly all their lives, and understood that they would continue to do so. Farming was the path given to them. Many of their lives were cut short by disease.

If an animal is healthy and eats well, it feels happy. But

man is different; he wants a purpose, he wants to leave a mark on his world. So I used to wonder if these laborers could be fully happy. Did their lives have enough purpose?

Now, as an old man, I see much better the advantages in earning one's bread by the sweat of one's brow. The peasants on my uncle's land knew that they must produce food. They were feeding the nation. The rhythms of the harvest absorbed most of their time and energy. Perhaps it also eased their cares.

But the extent of other people's happiness is always obscure, because happiness rests on experiences that are fundamentally private. Even the "progressive" landowner never really knows his peasants, no matter how assiduously he may visit them and consult their opinions on agriculture.

And what does happiness mean anyway? I have greatly enjoyed teaching, but on some days when the sun shone and a light breeze came through the window, I did not want to teach my class. Human desires are fickle and contradictory. Nobody enjoys all that they must do.

❧　　❧　　❧

When I was about 11, I was sent to bed for two weeks with a minor illness. A doctor came to the house and listened to my lung. After a time he told my parents that I might have tuberculosis. He was careful not to use the word "tuberculosis." He called it "a lung problem." But we all knew exactly what he meant. Tuberculosis killed many children in those days.

My mother asked Grandfather Einhorn to examine me. His diagnosis was more favorable; he doubted that I had "the lung disease." Still, he said, it would be wise to take measures to preempt it.

So my parents sent me away to a place of recovery, to

restore my health and discourage an attack of tuberculosis. It was decided that my sweet mother would accompany me.

When we left home, my sisters became quite emotional. My father said simply, "Look out for yourself. Be a good son. Be polite to your mother, and obey her carefully."

This place of recovery was high on an Austrian mountain, near the small town of Breitenstein. It was small and crowded with strangers. Most were wealthy adults, confined by their health to high altitudes. They lived as if in a hotel.

Most wealthy people then avoided speaking of illness. They used words like "sanitarium" to describe their accommodations though very few of us had brain trouble. They could not admit that human life is finite and often ends prematurely.

The natural setting was beautiful, but the human atmosphere was not. The sanitarium owners allowed us just two short walks per day. And you cannot walk far in 15 minutes. The food was good, but eating was no longer a pleasure. I passed many days, resting in one of those chairs where you almost lie. My mother reclined in the next chair.

Under other conditions, time spent in the mountains would have been a great pleasure. But the fear of tuberculosis had cost me my happiness. Even the presence of my mother could not cheer me.

I turned to mathematics. Sitting in my deck chair, I struggled to construct a triangle given only the lengths of the three altitudes. This is a very simple problem, which I can do now in my dreams. But then it took me several months of concentrated effort to solve it.

I knew that I might die soon. That was a very simple fact, but the meaning of it I could not fully grasp. I was told by my mother and by the hotel staff that by observing the hotel regimen, I would recover. I suppose that I should have feared the prospect of death, but I have never frightened easily. As long as they told me that I *could* recover, I felt that I would.

This attitude was a virtue, but it was not courage. To be truly courageous you must fully understand what is at stake. And at that age I did not. Dying was something abstract to me. I felt instinctively that I was far too young to die.

After about six weeks, the hotel doctors decided that my diagnosis had been mistaken. They said it carefully and with some reluctance, because a doctor hates to admit a mistaken diagnosis. But they said it clearly: I did not have tuberculosis.

So my mother and I gladly left this Austrian mountain, left the "hotel" there, the firm rules, the attentive staff, and all of the wealthy patrons there who dreamed of living forever.

I had learned that the human lifetime is finite.

# Three

# *A Tannery Needs Someone Who Knows the Work*

$\mathcal{M}$ ost women are practical. My mother, Erzsébet "Elza" Wigner, was no exception. She lived in the home, seeing that our food was well cooked, the house well cleaned, and the clothes well mended. She had mastered these tasks herself, but mostly she left them to the two unmarried servant girls who lived with us.

Keeping a smooth-running household was harder then. There were more children and they were more often sickly. A fire had to be lit and tended with every meal. Cleaning was done with rags and brooms. Washing was done by hand in tubs. So families with any money hired servants for these tasks. Our servants were almost like family members, but they took their meals alone. They were quiet and made no scandal.

My mother hired and directed the servants, as she did the governess. Keeping the house diverted my mother from many of the worldly concerns of my father. But she was kept busy loving and helping her family.

My mother was beautiful, but as a child I hardly noticed it. Her physical beauty played no part in our relation. I saw that she combed her brown hair to the two sides, and tied it neatly around her head. I knew that she wore high-heeled leather shoes; all our family had leather shoes because my father worked for a tannery. She never smoked, because it did not look right for a lady to smoke.

My mother came from a well-dressed family. Most of her skirts were white, yellow, or red, and her dressmaker made sure they were well cut. But fine clothes alone cannot give a woman beauty. Beauty comes from other sources. And she had those sources.

My mother kissed her three children good night, saying: "Ölellek csókollak"—"I embrace and kiss you." A common enough phrase in Hungarian, rarely taken literally. But my mother really did embrace and kiss us.

She had the skill of happiness and easily transmitted it to her children. She bought us rich chocolate embedded with candy and spoke sweetly. "Gyere az asztalhoz, itt az ebéd"— "Come to the table. The meal is ready."

My mother never openly corrected my father's appearance. But she always made sure that her children looked good. She spoke kindly, "Jenö, comb your hair again." And she helped my sisters to look attractive. Biri was married at 20 to an honest fellow, a minor executive for an important company. Manci was married soon after.

My mother was not an intellectual and never pretended to be. But she knew a great deal about what interested her. She knew the special weaknesses of her family: that I needed a great deal of sleep; that Biri could be too kind for her own good; and that Manci did not always like to play by the rules. My mother knew that her husband was a very private man, unwilling to discuss deep feelings. I think that she wanted him to speak of love and understood that he would not.

But, you see, it was her duty to understand. The wife was subordinate to the husband in those days. It was my father who bought our summer house 20 kilometers north of Budapest and even decided when we should use it. My mother was thought to have a lucky and successful marriage because her husband was well-to-do, truthful, and decisive.

She never disputed my father in our presence. Arguing and embracing were two things that were simply not done in front of children. She relied on my father's income, knowing very little about how it was acquired. Her job was to spend it wisely, and she did that well.

My mother kept close to her four siblings. Her brother lived in Budapest and visited us nearly weekly. Then she had three younger sisters: Frieda, Otti, and Margit. The fullest kind of beauty combines physical beauty with beauty of temperament, beauty of spirit. Frieda had that fullest kind of beauty.

One of my mother's sisters married a Budapest physician, and their family often visited us. The other two sisters lived as country landowners, the one in Belcsa-Puszta, in southern Hungary, and one near Győr, west of Budapest. My mother spent much of her time keeping up with her extended family.

She talked freely about what was cooking for dinner or whether we should take a walk. I rarely knew just what went on inside her; in that way, she was quite private. But in daily affairs, she was close, helpful, and dear.

My father, Antal Wigner, was altogether different. "Anthony Wigner," he would be called in English. My mother called him "Toni." "Wigner" had once been the German name Wiegner, meaning "cradle maker." I was told it was my great-grandfather who had made it "Wigner."

My father was born in 1870, so he was about 38 when I first knew him. He had brown hair and gray eyes, and was considered a tall man at 5 feet 10 inches. He was an only child born to a leather-tanning family. He grew up in the town of Kiskunfelegyháza, 125 kilometers southeast of Budapest, not far from the city of Szeged.

Before my father was 2, his father had died, and he had come to Budapest with his mother. Somehow she knew the owner of a great Budapest tannery, Mauthner Testvérek Astársai—Mauthner Brothers & Collaborators.

The Mauthner Brothers tannery, the second largest tannery in Hungary, was then already about 150 years old. It had stayed in the Mauthner family, prospered, and grown. Leather-tanning methods had greatly improved. Finished leather was sold for shoes, saddles, satchels, and leather coats.

Young Antal went to work there right off, and slowly worked his way up. All day he worked at the tannery and in the evening attended the Lutheran school, or gimnázium. He knew that if he misbehaved even slightly, his mother would be called in. Always he behaved.

When my father was about 16, his mother died. My father was then quite alone in the world and obliged to work to support himself. And this he did. If he had ever hoped to attend university, to study physics, or become a scientist, he gave that up. My father had been an orphan for almost 15 years when he married my mother, and in those years had grown quite independent.

If he dreamed about a fuller kind of life, he never shared his dreams. If he had known sisters or brothers, if his parents had survived, he might have been gayer, more animated. But such things had not occurred. So while my father was devoted to his wife and children, his family life was quite constrained.

He liked a good joke, but rarely told jokes himself. He had no interest in running the household. When he played cards, it was solitaire. When he came home just before dinner, he did not say: "Hello, my family. I am home! Come here, let me greet you!" No, he just entered the house, making clear his return. He never embraced or kissed his children.

My father liked my mother's family, but far less intensely than she did or his children did. I do not know how my parents decided to marry. They may not even have made the choice; perhaps relatives persuaded them to unite. I never knew and it was clear that I should not ask.

My father was not physically active. He took walks and perhaps a few scattered hunting trips. But he never rode a bicycle and rarely worked in the garden.

He dressed well, usually in a dark suit with trousers, vest, coat, and necktie. Heat or an informal setting might persuade him to remove the jacket, but the necktie stayed on until he retired for the evening.

It seemed to me as a child that my father had begun his life at the Mauthner Brothers tannery. That was the first part of his life he was willing to recall. He told us how the tannery owners had favored him, gradually made him a director, and had even given him a financial share in the business.

Antal managed about 400 employees. He helped decide which materials the tannery should buy and from whom, how the work should be organized, and what products should be advertised. He did this quite well, without ever being a dictator.

The tannery was located in Ujpest ("New" Pest) about five miles north of Budapest, east of the Danube. Tanning was done on a grand scale, in a process with many stages, requiring a building larger than a city block. The land required was far cheaper beyond the city.

I could still find that building on the street. Inside, it smelled of tanning chemicals. There, the hides and skins of dead cows and bulls were cured in large vats. They were trimmed, sorted, washed, and treated with tanning acids, then bleached and coated with oils until they were firm and resisted water.

Powerful machinery was used, but most of the work was done by hand. I recall one fatality, but the work was generally quite pleasant. There was little mechanical noise; just the sound of workmen speaking Hungarian. Their clothes were protected from the tanning acids by a heavy smock.

A horse-drawn carriage came each morning at 7:30 to bring my father to work. Later, the horse and carriage were replaced by a tannery automobile, a Benz I think. My father never learned to drive a car. He rode to work next to the tannery owner. After a few hours in Ujpest, the driver took my father and some others to the smaller main offices in downtown Pest. Their office work was assisted by female typists.

My father might have failed as a doctor or teacher. But he would have graced nearly any other profession. He was an able protagonist, a skilled handler of people and events. These facts were well known in the society he kept.

My father was very responsible. He liked Hungarian cigarettes, but he never smoked in the tannery's industrial building because there was a fire regulation against it. He was always proud of being in control. I doubt he was ever drunk in his life.

But my father was not overly proud. He treated the largest tannery in Hungary as his rival; he knew that his own tannery could not easily surpass it, and that doing so might even disgrace the other tannery. He did not want that. He wanted a world of reason and order.

Sadly, many workers in the tannery disliked my father. He was a factory boss at a time when such bosses were powerful and rarely popular. Just as my mother ran an efficient household, my father liked to run an efficient tannery. If it was not a dictatorship, it was not a democracy either.

When I joined the tannery, I cringed to see clearly the workers' dislike of my father. When you like your boss, you find time to idle with him and share some small talk. There was no small talk between Antal Wigner and his workers. They avoided him.

But I admired my father greatly. He was observant. He rebuked tannery colleagues who used poor methods and family members who asked to borrow his money. He spoke plainly and pointedly, but never violently. And his criticism was helpful and fair.

At heart, my father never disapproved of me. He often scolded me a bit for stupidity. He also told me seriously how I could improve myself. Sometimes at home he would say, "Do not hurt Jenö, especially not in his head. For that is the weakest part of his body!" That kind of gentle teasing was typical of him.

I never teased my father; one does not tease the boss. I did not start private, unnecessary conversations with him. You know, I am sorry that my father is no longer living. But if he were, I *still* would not ask him unnecessary questions. I might ask when he would like to eat and when we should attend the theater. But I would not ask him personal questions.

I scarcely had famous heroes as a child. My feelings for God were neither warm nor intense. I never had the usual boyhood idols of artists, scientists, athletes, or politicians. I did not follow the serial exploits of boy heroes in cheap novels. My parents were my heroes, especially my father.

At 17, I began asking my father some of the questions that boys near manhood have always asked their fathers: How great is the world? How much can anyone know of it? And how much *should* one know?

My father was an intelligent man who instinctively knew a great deal about the workings of human society. Very little escaped his notice. But he was also a very formal person, devoted to a job that was highly specialized.

As a result, my father felt that he had little time to consider such things as how great the world is and how much any one man should know of it. I learned a lesson from my father's attitude, and I have always taken care to avoid highly specialized work, so that I can keep an interest in those great questions I first pondered at 17.

My parents had created me, and that fact gave my father an overwhelming psychological advantage in our discussions. But despite the awe I felt for him, I could not help but disagree with my father on some of the great questions. I was aware even then that he ought not to determine all of my opinions.

So my father and I argued over religious questions, over the characters of our friends and of government figures, and over the behavior of past kings. My father held that human desires are not grounded in reason. He did not entirely trust human nature.

I argued that the great majority of people deserve our trust. But though I could dispute my father's logic, I had to respect his person and the power that he projected. He was a kind man, but he knew how to make his authority felt. Antal Wigner won nearly all of our arguments.

Once I asked my father, "Why are people so attached to

money?" He responded simply, "Because of the power and influence it gives them." I disliked this bit of cynicism and told him so. It was years before I saw that he was largely right: The human desires for power and influence are very deep and strong. I learned a great deal from my father which I failed to fully credit at the time.

These talks with my father led me to wonder, "Why am I on this earth? What do I want to achieve?" I felt my purpose should be to marry, to begin my own family, and to provide this family with a proper home and nourishment. Today, these things come far more easily and many youths no longer know what to strive for. Many of them see power and influence as the only valid goal. But in 1919, providing a home and nourishment was a valid purpose.

❊      ❊      ❊

I never knew for sure that my father's family was Jewish. The only thing that I knew definitely about them was that they were no longer alive. My mother must have been of Jewish descent, but I never asked her about it. I knew it was an awkward subject for my parents. At rare times, walking alone with my father, I asked him—implicitly—if his family was Jewish. He suggested that they were. But I never asked directly and he never told me. Talking intimately was not his habit.

I think we can assume my father was of Jewish heritage. If he had not been, he would not likely have married my mother. In 1900, Jews and Christians very rarely married outside their faith. My father probably told my mother his religion, but even that I do not know. Only after leaving home did I see how many things my parents had kept from me.

We had a little Seder when I was a child and I was prepared for the Bar Mitzvah, because that was the Jewish way of becoming a man. But my Bar Mitzvah had no religious meaning to me or to my family. My best Bar Mitzvah gift was a pocket watch. I got it on November 17, 1915, and I still use it today. I have to wind it daily. But it has not needed repairs since 1921.

I was enrolled in the same Lutheran gimnázium that my father had once attended. I was a Jew in religion class, but hardly in spirit. I went to school on Saturdays and very rarely wore the yarmulke. I never understood the reasons for circumcision. We never spoke Hebrew at home, never had Shabbat dinner on Friday, or kept kosher. We had a Hungarian Bible, but rarely consulted it. We went to synagogue perhaps twice a year. Religion was a painful issue for the Wigners.

So we did not practice Judaism, but stayed nominally Jewish. I never thought to ask my father why. I felt we were Jews because our ancestors were Jews. In the days before the First World War, people thought no more of changing religions than of changing their sex. To change would have been a swindle.

The First World War was a bitter thing. It changed the Hungarian borders and hurt our national spirit. It began in June 1914, when Archduke Franz Ferdinand was assassinated, reviving in Vienna the fear of rebellion against the Austro-Hungarian empire.

The First World War may not have been an inevitable war, like the East Asian conquest of Hungary in 1896. But neither was it started by a ruthless dictator. And Hungary's role began modestly, one month after the assassination, when it joined forces with Germany and Austria. We were told that it was a defensive battle only, directed only against our southern neighbor, Serbia. Why did this little war spread into the massive slaughter that engulfed much of Europe for four years? That is a question for the historians.

What I recall is the simple faith with which we entered the

war. We were told that the war was caused by Serbia and expanded by Germany. But our leaders told us that the cause of the war hardly mattered, because the war would be decided within three weeks. By then, our powerful new rifles and machine guns would have killed or wounded nearly all the ready soldiers on the other side and forced a surrender. But a series of treaties quickly drew many other nations into the war on both sides. Neither side surrendered; instead, they dug trenches. And this small defensive war became the terrible "world" war chronicled in the history books. The Germans, Hungarians, and Austrians lost the war. Especially on the Italian and Russian fronts, hundreds of thousands of Hungarians died.

My parents neither approved nor disapproved of the war. They knew it had been started by governments, would be managed by governments, and finally halted by governments. Approval or disapproval from the Wigners changed nothing.

So my parents hardly criticized the conduct of the war. We were not the sort of family to critique things we could not change. Quietly, we hoped for a Hungarian victory. But we knew that nations are usually foolish to seek new territory. Near the end of the war, we feared that Budapest would be destroyed. It did not happen, but the mere risk of it showed people how helpless they were to resist forces they scarcely understood.

My father was too old to be drafted. I was too young. But by the end of the war I was 16, and if it had gone another year, I might have been called. That worried me; I would have been a reluctant soldier and a poor one.

Finally, the war ended. Hungarians were relieved but subdued as well. It was a very harsh defeat for the Hungarian side. The Austro-Hungarian empire was split up and gone. Hungary was again a republic and an independent nation.

There was some justice in this; Austria–Hungary had always been run from Vienna, with Hungary a quite junior partner. Many Hungarians had resented this, feeling a distinct difference between ourselves and the Austrians. We called ourselves "Magyars," and this word had magical properties.

My family carefully stayed neutral on the nationality question. We advocated neither subservience to Austria nor an aggressive national independence movement. We advocated being helpful and not starting any further wars.

The new Hungary was far smaller than the Austro-Hungarian empire, smaller even than old Hungary. The war swept away a great many hopes. Many Magyars found themselves living in the new state of Rumania, and even today they struggle to form a separate Magyar government there.

For many Jewish families, the nationality question was tied up with bitter memories of anti-Semitism. But not for the Wigners. I almost never felt anti-Semitism personally in Hungary, though I saw how it was used to control the prospects of the Jews.

For many years, nearly all Hungarian Jews had been merchants. Gradually, they had moved into the professions: law, medicine, engineering, the sciences, and teaching. But they had remained under severe restrictions. No Jews were admitted to the national military academy. There was a Jewish membership quota of about one tenth in Hungarian universities. A Jew had no hope of receiving a high government post; most Hungarians considered being ruled by a Jew to be a great indignity.

But very few Jews ever protested this state of affairs. Certainly, we felt that in a perfect world the Jews would have some political power. But we expected to see that no more than we expected wisdom to become universal or all disease eradicated.

As Jews, we knew that if we were lucky and bright we

could support those in power in a number of intriguing ways. We could lend money or, even better, our political counsel. A Jew with political ambitions beyond the role of trusted advisor was often thought conceited, even by his fellow Jews.

It is hard to believe today how strict the barriers were against Jews in the official life of the nation. But in those days society was far stricter generally. Nearly every part of life was strictly governed.

Jews were citizens of many countries, and Gentiles felt that Jews loved Judaism far more than they loved their country. So, they reasoned, Jews should not be entrusted with the most important jobs, with no shortage of qualified "real" Hungarians eager to fill them.

But even these "real" Hungarians had far less influence than a common citizen has in the United States. Far above both Jew and Gentile in Hungarian society was the Austro-Hungarian royalty. The great understanding that shaped the lives of Jews and Christians alike was that our country was owned by a mighty king, and to a lesser degree by his wife, his friends, and his children.

While holding this bounty, our king was obliged to lead and inspire us and to protect the nation from both internal decay and external invasion. But he was never obliged to inform us just how he did this. He was expected neither to publish his desires nor to justify his actions. He simply acted. The Hungarian king was far, far above his subjects, somewhere up in the clouds.

In school, some groups disliked Jews. Various anti-Semitic incidents occurred, as they have occurred in every nation throughout history. Man lives uneasily with those of other religions. When such incidents touched my own life, I tried hard to forget the details, and with age I have succeeded in this completely.

I do know that the Wigner family was content enough as Jews. One biography of Edward Teller says that I was "beaten by a mob," probably for being Jewish. Well, when I was 16, I was in a fight, yes. I do not remember it well, nor do I want to. But this "mob" was no more than three. Such hoodlums do not bother to define the source of their anger before assaulting their victims. But I doubt it was my Judaism.

Far more troubling to my family than anti-Semitism was communism. Sadly, the two were related. About 1915, the communists began gaining real strength in Hungary. My father deeply opposed them. He felt that they would cruelly restrict our lives.

Many of the top communist leaders were Jewish, and my father found this quite disturbing. He was only embarrassed about it at first, but as Jews became more strongly associated with communism, my father took a radical step: He arranged the conversion of his family to Christianity.

My father only considered Protestant denominations. Roman Catholicism seemed too much like communism, a well-run dictatorship. My father had been to the Lutheran gimnázium and had enrolled me there too. So it was natural for him to pick the Lutherans.

My father discussed this decision with us. My mother likely had a chance to disagree, but she did not protest our conversion in any way that I could see. My sisters and I had no objection. So we all became Lutherans. The Lutheran pastor gave warm, lively sermons, assuring us of God's existence and instructing us on how to enjoy life.

I think my father enjoyed Lutheranism, but our conversion was not at heart a religious decision but an anti-communist one. Since the first World War, Jewish conversion to Christianity had become far more respectable. Jancsi von Neumann's family became Roman Catholic; apparently

Catholicism did not much remind Max von Neumann of a dictatorship.

I have never missed Judaism. But I consider its beliefs little different from Lutheranism. The fundamental law of all the major religions is be kind, cooperative, and helpful. The important questions are: Should you help sinners? If so, which sinners, and in which ways? Should you have one wife or two? On these questions there is little difference among the major Western religions.

Theological disputes about the proper role and character of the church, the ruler, and God—these things do not touch our daily lives. Belief in a threefold or a singlefold God does not change your activities. Christians believe that Jesus Christ is their personal savior and Jews do not. But what does this matter in daily life?

So the Wigners easily became Lutherans. Today, I am only mildly religious. When I attend church, it is with the Protestants. My wife and children are not Jewish, and I know few Jewish people well.

Like my father, I opposed communism in 1918, feeling that Hungarian life must not be ruled by the government. We had no communist writings at home and they certainly were not in the gimnázium curriculum. But they were discussed socially at school, and I was curious enough to locate some in a library.

I found Karl Marx's work quite unconvincing. And Lenin was even worse. That all power should rest in the state was an idea with quite obvious and serious flaws. Lenin's work clearly brimmed with a lust for power and a grotesque urge to regulate human life by the tenets of communist ideology. Rigid organization may create a perfect antheap, but even at 16 I knew that human beings are not ants and need far broader freedoms.

This was before most of communism's worst crimes. Some people have asked me if the intensity of my anticommunism is an emotional reaction to childhood events. It must be so, in part. I saw the crimes of communism.

But I have opposed communism for rational reasons. Communists insist on great personal power and are always tempted to try extending it. They begin by telling you what to do and how to do it. Eventually, they want to tell you who to love and what kind of children to produce.

So communism rarely brings the happiness it promises, but instead a general feeling of subversion. The world now has 70 years' experience with communism, and I am often surprised how many people still believe that communism gives greater social equality than does democracy. It gives less. Communist governments reward their friends and punish their enemies far more than democracies.

Why did the Hungarians let the communists seize Hungary from the old monarchy? Well, the change came in stages. First came a leftist revolution against the Hungarian monarchy in October 1918. The leftists, led by a Hungarian count named Károlyi, threw out the monarchy and promised that Hungary would separate from Austria and enact various social reforms. In the process, the leftists also gave much of the communist platform a legitimacy it had never enjoyed before.

And Hungary still had economic and political problems that the leftists clearly could not solve. Now who would solve them? The communists stepped up and promised to do so. In the midst of our bitter national confusion, the communists cleverly proposed simple solutions: Redistribute land! Seize the organizations! Change the rulers!

After just five months, the leftist government fell to the communists in March 1919. The top communist, Béla Kun, was the son of a Jewish village clerk. Kun was not much over 30, but he was brash and strong. Most Hungarians were prepared to accept his government, at least for a time.

Many people, especially the young, felt that communism would make them as rich as landowners, as strong as factory directors. Wealth and a life without landowners! With these slogans, communism claimed to foresee the future.

Three communists were in my high school class. They sincerely believed that capitalism and monarchy had poorly served our country. "Look here," they would say. "Our factory directors are chosen unfairly. They have a father or uncle already installed in such a position. He arranges their appointment!"

My three communist classmates felt that the communists would end this kind of favoritism. They would make appointments by the "objective" new communist system. And in this and many other ways, communism would elevate the people.

I opposed these assertions, but my objections went unheeded. And as communist forces gained strength, these classmates assumed a certain prestige. I doubt they stayed communists during the whole regime and its overthrow. Most of the early communists did not.

As rulers, the communists were hardly cruel, but they were quite unreasonable. They did not help all the people, as they had promised. Their government seized some land and power. But power in government hands rarely helps the people.

The communist system also fueled antagonism between grades of worker. Sickly workers could be fired, but workers could no longer leave one factory for another. Every worker needs that freedom, even if he does not use it. The mere knowledge that he has it is invigorating.

And the top communist rulers did not make "objective" appointments. Not at all. They often made their friends subrulers, and many of these subrulers began reforming departments that they did not understand. Naturally, their subordinates resented it.

The communist ruler appointed to the Mauthner Brothers tannery was ignorant and foolish. The heads of the tannery were thrown out, my father was thrown out, and the tannery was taken away.

A tannery needs a leader who knows the work and can decide: "Fifteen men will work on this, twelve men on that." Even the communists knew that this leader should come from within the tannery. So perhaps the new communist tannery would not have been so different from the old one. Perhaps my father might have soon gotten his job back. Let us hope so.

But the Wigners did not stay to find out. Soon after the communists took over Hungary in March 1919, my father moved his family to a summer resort in Austria, south of Vienna. I doubt my father thought of any other haven but Austria. It was near and both my parents had relatives there. Many Hungarians left Hungary then, including the von Neumanns.

None of the Wigners were very happy in Austria. This uncertain period is one that I have tried to forget. We lived uneasily in the resort. I liked learning more German, but we knew that this move could not last. My father had very little money. He had a distant cousin Wigner in Vienna, who owed him money and paid some of it. But my father was out of work.

I first broached with him the subject of my career plans during this unhappy stay. "Father," I said, "if the Hungarian government will not allow anticommunists in the tannery, there isn't any sense in my learning tanning."

We had an awkward discussion. I knew my father resented the interruption of his career. I hated to inflame his feelings, but what I said was true. And I knew that I would prefer to be a teacher or scientist than a tannery executive. My father said, "Well, if this is the world we must live in, perhaps you are right. But don't decide yet."

All of the Wigners knew that the cause of my father's unhappiness would not be discussed and that we would all return some day to Hungary, the only place where he could properly support us. But these things were never spoken aloud.

Communist rule of Hungary lasted only from March until November 1919. In November, the communists were overthrown. Soon a conservative Hungarian admiral named Nicholas Horthy was installed as a regent, acting in the place of a Hungarian sovereign.

My family had feared a far longer period of communist rule. When Horthy took power, we joyfully returned home. My father returned to the tannery, his wife to the home, and his children to school, just as if communism had never occurred.

The new regime turned out to be bitterly, unreasonably anticommunist. They executed about 5000 accused communists and jailed about 70,000 more. About 100,000 Hungarians were exiled. Because Jews were associated with communism, Jewish rights were further curtailed. Jews were very nearly barred from Hungarian universities entirely. That was disquieting.

But our primary feelings in 1919 were joy and relief at the fall of the communists. We knew scarcely anyone who had been jailed, executed, or exiled. Perhaps we should have done something for them, but we did not. We had now seen dictatorships of both the Left and Right, and heartily disliked them both. But we were deeply glad to return home.

## ∾ Four ∾

# "How Many Such Jobs Exist in Our Country?"

*I* feel instinctively that the great majority of human talents are inherited and the learned portion very slight. But I cannot gauge the truth in this, and much of the experience of my youth seems to contradict it. A series of mentors in those years showed me just how far even an ordinary student can go when he is brilliantly taught, blessed with luck, and deeply in earnest.

Budapest in 1915 was filled with fine high schools, including the Minta or "model" gimnázium attended by Leo Szilard and a few years later by Edward Teller. But on a wide street called Fasor was located my Lutheran gimnázium. It was likely the best high school in Hungary; it may have been the finest in the world.

Few Hungarian boys then finished high school. Eight years of school satisfied all but the professional families. The few boys who went on to gimnázium gained from its intimacy in two basic ways.

First, we knew each other thoroughly. I never had to ask a boy his name. I knew him. I walked home from school with him or with his friends. I knew where he lived and what his father did for a living. I knew what he liked and disliked, what order of sums he could perform in his head, and where he would need a tablet. I knew what humored him. And 50 years later, I would still recall our student days with real affection.

The second virtue of a small gimnázium was that few teachers were needed and they could be chosen with great care. Our teachers were superb. Several of them did independent research, but most enjoyed teaching more than anything else.

Our gimnázium teachers had a vital presence. To kindle interest and spread knowledge among the young—this was what they truly loved. They were preoccupied with teaching and they impressed us all not only with their array of facts but with the intense and loving attitude they held toward knowledge.

Gimnázium teachers ate lunch together. On Saturday afternoons they often met at a coffeehouse to discuss their work with university colleagues. I was invited to join a few of these meetings. In American high schools, such social mingling of students and teachers rarely seems to occur. Hungarian teachers watched their best students closely.

So we learned a fantastic amount at the Lutheran gimnázium. Six days a week we attended, and no one grumbled at that. We did not just memorize facts coldly; we put them into our heart.

Diligence and mental power were widely honored. I tried to read ahead of my class. A few students teased me gently about my deep love of mathematics. But such depth of feeling was respected by my peers. A good many of them shared it.

The Lutheran gimnázium was affiliated with the Lutheran church, and the school building was literally connected to the church. Several gimnázium teachers were on a committee advising the church. Most of the school's prominent teachers were Protestant. But in our weekly period of religious instruction, a local priest and rabbi were brought in to instruct boys in their respective faiths.

I went first to Jewish class. I must have done very poorly. The same young rabbi came in each week. He told us that there was a Lord, that He was all-powerful, and that we should submit to Him. Everything in the universe was apparently the creation of this one deity.

To this assertion I could not help wondering: "Everything?" I felt some natural affinity with religion. I admired its lack of self-consciousness about fundamental questions. But the idea that every single thing in all the universe had been created by one deity did not get very far with me.

I sensed at an early age that all the world's major religions felt awkward around science. Both religion and science were trying mightily to explain the great wonders of the world. Each seemed jealous of the allegiance shown its rival. But science seemed to me less jealous and far closer to the truth. Very early, I cast my allegiance with science.

The rabbi who appeared at gimnázium never knew that, but I doubt that he would have reacted strongly if he had. He did not doubt his work. He led us firmly through Bible readings in both Hebrew and Hungarian and taught us Hebrew grammar and vocabulary until we could read some fine Hebrew texts. But we used a kind of baby Hebrew, with vowels. I never spoke or read formal Hebrew with any skill.

My heart was with numbers, not words. After a few years in the gimnázium I noticed what mathematicians call the Rule of Fifth Powers: That the fifth power of any one-digit number ends with that same number. Thus, 2 to the fifth power is 32, 3 to the fifth power is 243, and so on. At first I had no idea that this phenomenon was called the Rule of Fifth Powers; nor could I see why it should be true. But I saw that it was true, and I was enchanted.

I demonstrated this phenomenon to my father. He was pleased and amused by it, as anyone is pleased to see a pattern growing out of the air. But my father had no more idea than I did of why this pattern should exist. Indeed, we both looked on it as a kind of trick.

One of my uncles took an interest in mathematics. When I shared the trick with him, he was less impressed than my father and told me that the secret was contained in Pascal's triangle. This was entirely wrong, but on my uncle's word I blithely accepted it for some years.

My family made only perfunctory attempts to learn why I pursued such abstractions. "He's a queer child," they said. "But it doesn't matter much."

❧          ❧          ❧

As a child I knew almost nothing about the United States. Gimnázium taught us little more about this country than that it was one of the Americas. We studied North America and South America; the narrow Panama Canal; Mexico, Brazil, Chile, Argentina, and Canada. But none of it seemed real. We had never seen these countries and never expected to. We read about them as we read about the Roman Empire.

The phrase "the United States" was translated for us literally into Hungarian. I had trouble grasping the concept. How could independent states be united? That the United States had made itself a democracy was hardly mentioned. We learned something of America's successful revolution against the British. That, too, puzzled me, because we had been told that Britain had ruled the earth in 1776. So how had the Americans defeated them in a major war?

The gimnázium gave grades one, two, three, and four. One was best, four worst. Only about four boys regularly got grade one and we knew each other quite well. A few students were far weaker, but that could not be helped. I studied no more than three hours a day. My parents knew that I did the best I could easily do and accepted my grades and study habits without comment.

There were many superb teachers at the Lutheran gimnázium. But the greatest was my mathematics teacher, László Rátz. Rátz was known not only throughout our gimnázium but also by the church and government hierarchy and among many of the teachers in the country schools.

My first meeting with Rátz deeply impressed me. I had come down with typhoid fever and missed four months of school. Even after recovering, I felt awfully weak. After missing so much school, I had to take a special examination in the fall. Rátz was my examiner. Such examinations were as unpleasant as the examiner wished them to be. But Rátz made the test pleasant. He asked a great many questions, but his interest was entirely friendly. I saw vividly his passion for mathematics.

Rátz's family had originally emigrated to Hungary from Turkey during the Turkish occupation of the seventeenth century. Rátz was a tall, blond man. Perhaps to compensate for a balding head, he wore a thriving mustache. Rátz was about my father's age but seemed younger. He moved with an athlete's grace. Rátz was unmarried, though not entirely innocent. He had been teaching at the gimnázium since about 1890.

I still keep a photograph of Rátz in my workroom because he had every quality of a miraculous teacher: He loved teaching; he knew his subject and how to kindle interest in it. He imparted the very deepest understanding. Many gimnázium teachers had great skill, but no one could evoke the beauty of a subject like Rátz.

Rátz cared deeply about mathematics as a discipline. He had founded a mathematical journal for secondary schools which came to be read all over the nation. He ran the journal for 20 years, often distributing it with his own money.

At the start of every month, Rátz collected a sizable sum of money from nearly every faculty member, and placed this money in a fund designed to ensure the material comfort of the teaching staff in their retirement.

Rátz was also something of an educational reformer. Around 1900, he had pioneered a new and effective method of teaching mathematics. Though the method was not yet officially sanctioned by the government and clergy when I was at gimnázium, it was clear that it soon would be.

But Rátz devoted most of his genius not to broad reforms but to the act of teaching mathematics itself. At the retirement of Imre Gobi, the gimnázium director, the staff named Rátz as his successor. They gave Rátz a formal title and likely a higher salary. Most men would have said, "Thank you kindly for the promotion. This is very fine." But Rátz worried that his new duties would hurt his teaching. He knew how much energy is needed to evoke the deepest beauties of mathematics.

And after five years of distinguished service as director, Rátz decided that the extra work had indeed hurt his teaching. He quietly resigned as director and became just a teacher again. Some staff members were bewildered by his resignation from such a prestigious post. But Rátz cared little for prestige; he was a remarkably unselfish man.

A pamphlet written around 1925 honoring him on his retirement from the school described Rátz as "supporting the

weak, correcting the wayward, and strengthening the faint."
All those things he did. But Rátz also took special care to find
his better students and to inspire them.

The great mathematician Johnny von Neumann was
Rátz's best student. Von Neumann, then known as Jancsi,
was a year behind me at the gimnázium. Von Neumann was
truly a prodigy and already had made mathematics his first
love. To be a mathematics prodigy takes not only great talent
but unusual devotion. When Rátz saw how intelligent Jancsi
was and how devoted to mathematics, he began giving him
private lessons.

You might say, "Well, von Neumann was one of the
great mathematicians of our century. Of course he deserved
private classes as a boy." But look at this from Rátz's point of
view. You have very few students. One boy, von Neumann,
soars above the rest. He appears to be a genius. But, of course,
he is not yet famous at all. His brain is not adult. He has never
published; he has not yet even produced anything original. He
is just a startling 10-year-old boy, working right next to 20
other bright 10-year-olds and rapidly mastering much of what
is known in mathematics.

How can you know that this precocious 10-year-old will
someday become a great mathematician? You really cannot.
Yet somehow Rátz did know this. And he knew it very
quickly.

Rátz felt so privileged to tutor a phenomenon like Jancsi
that he refused any money for it. Jancsi's father, Max von
Neumann, was a banker. In 1913, he was in the process of
being ennobled by the Hungarian royalty. That is how the
"von" came into "von Neumann"—for Max Neumann's patriotic cash contributions to the Austro-Hungarian nation.

So Max Neumann could afford to pay handsomely for
his son's tutor. But Rátz did not want money for teaching

Jancsi von Neumann. His compensation was subtler: the brush with a special kind of mind; the privilege of training that mind in a discipline that both of them loved.

Most of the parents of Rátz's students were quite fond of Rátz and many of them consulted him about their sons' upbringing and choice of profession. So Rátz likely visited the von Neumann home to ensure that the von Neumanns understood the nature and implications of their son's gift. He probably asked them to help Jancsi become a professor or scientist. Rátz even put von Neumann in some classes at the University of Budapest. I never knew another gimnázium student to take university classes.

Rátz was just as nice to me and nearly as devoted as he was to von Neumann. There were no private lessons; I never expected them. But Rátz lent me many well-chosen books, which I read thoroughly and made sure to return in good condition.

One of these was Hesse's *Analytic Geometry.* There was also a book on infinitesimal calculus; I hardly understood it, though I thought I did. A third book described the Small Theorem of the seventeenth-century French mathematician, Pierre de Fermat. Fermat was a contemporary of René Descartes and a founder of the modern theory of numbers. It dawned on me that with Fermat's small theorem I could easily derive the rule of fifth powers which had so puzzled me a few years before.

Rátz also gave me books on geometry; books on the drawing of functions in a graph; books on ordinates and abscissas, circles, and ellipses; books on first- and second-order equations; and even books on statistical mechanics, which then was not yet quantum mechanical. I filled a small journal with notes on what I had learned. I loved especially a three-volume mathematics text by Hans von Mangoldt.

Recognizing the laws of geometry, reading mathematical articles in German, discovering a proof of the simple theorem that the three altitudes of a triangle meet at a common point —these things delighted me.

There was just one boy in my class who loved mathematics as much as I did. His name was Riemer and we often worked together. Riemer was a political leftist, but affable and kind. Rátz gave the two of us the hardest problems during examinations.

Rátz also compiled for his students a book of "commonsense" mathematical problems. I solved a few of them, but most I found fantastically hard. Often in the years since, when I have been in no mood for work, I have taken Rátz's little book from the shelf and studied those "commonsense" problems. They are grown-up problems.

Rátz was the only gimnázium teacher to invite me into his home. He would ask me informally to spend an hour with him in the afternoon. I used to bring him little gifts. We would sit drinking coffee and discussing mathematics. We hardly ever discussed the nature of our relation or what these afternoon visits meant to us. But we knew very well how we felt about each other.

I could not see then the full scope of Rátz's influence on me. Some things one only realizes full-scale years later.

Let me introduce some of my other fine teachers. Ferenc Szolár gave us a daily Latin hour. Hungarian law required six hours of Latin a week. We learned Latin grammar and read some of the famous Latin writers. Szolár asked good questions and explained things so that they were understood. Thanks to him, I can still read and translate Latin, though it does not excite me.

Jenö Gretzmacher taught mostly German language and literature, including the story of Wilhelm Tell and the apple.

Imre Opel, a young man, gave us the rudiments of geometry and physical education. Somehow I learned a bit of French at gimnázium too. At $14\frac{1}{2}$, I chose to study art. If I had not, I would now be able to read and translate Greek.

I also learned poetry at school. Three nineteenth-century Hungarian poets marked me deeply: János Arany, whose poetry described deep and lasting emotions; Sándor Petöfi, who wrote quick, lyrical poems and died at 26 in 1849, during the Hungarian Revolution; and Mihály Vörösmarty, who wrote sweet, expressive poems of patriotic sentiment.

Poems are wonderful. You cannot appreciate them in a literal way. But you feel them. They come to you and grow in you. With very little discussion of this poetry, I developed a violent emotion for it before I was 20. I have never lost it.

Latin, Hungarian language and literature, German language and literature, mathematics, history, religion—all of these subjects were taught seriously. Science was taught less seriously, and this irked me. Physics and chemistry were hardly treated more seriously than geography or art. We had just one year of chemistry and two of physics.

We did learn some of the history of science. We studied the Egyptians and how they measured the inundations of the Nile; the Babylonians, who discovered the scientific method about 900 years later; Archimedes, who built a fine catapult and found that objects are lighter under water; Plato and Aristotle, who first divided the mind into intelligence, feeling, and will.

The physics teacher at the gimnázium was Sándor Mikola. Mikola seemed at first a gruff, distant man. He came, imparted his wisdom, and left. But he had a good heart. He was quite plump, and amused us unconsciously by playing with his pince-nez. But though we smiled at his foibles, we respected him.

I already loved physics and I wished that Mikola taught with the vigor of Mr. Rátz. But Mikola knew physics well. He had written a fine physics text. He told us about Sir Isaac Newton and his equation for the motion of stars and planets; of Johannes Kepler and his astronomic physics; of James Clerk Maxwell and electromagnetic theory. He taught us that electric current flowing into a coil of wire induces a magnetic field and showed us how to find the direction of the field. He spoke clearly and well and conducted sound experiments.

I did some physics independently. I proved the energy theorem to myself and was quite pleased by it. We had a small student colloquium at the gimnázium. It was voluntary, but nearly everyone joined. I gave a short lecture to this group on relativity theory. Afterward, Mikola called me in for a talk. It may have been the only time I ever spoke with him at length outside class.

Mikola asked me bluntly why I believed in relativity theory. I described with some animation the implications of the Michelson–Morley experiments. These experiments, organized in the United States by the Prussian-born physicist A. A. Michelson, had used an optical device to establish the velocity of the earth. The results had implicitly contradicted the physics of Isaac Newton and had helped lead Albert Einstein to his remarkable theory of relativity, which he had published back in 1905.

Mikola did not like being reminded of all that. But he accepted my enthusiasm with the comment, "Well, the grass is always greener in the next pasture." Mikola treated relativity theory as a novelty and asked me never to forget how many vital problems were beyond its reach. I promised that I would not.

The physics text that Mikola had written was my first real physics book. It told us: Atoms and molecules may exist, but the question is irrelevant to physics. Physics is entirely concerned with things which can be observed. That was a nineteenth-century view, but it still had a great many adherents in 1919.

I felt somehow that Mikola's book was wrong to deny atoms and molecules a place in physics. I took offense. I felt the lack of atoms and molecules not quite consciously, but at some deeper level, as millions of children in Communist countries have grown up sensing the existence of a freedom that is denied them. My curiosity for atoms and molecules led me to read other physics texts, some of them fairly advanced.

Mikola knew a great deal of physics. Certainly, he knew enough to see that the physics he taught was incomplete. But he did not see an opening to explain the rest of physics and he accepted that lack, as even an intelligent city dweller is apt to accept his ignorance about much of the city he lives in. Imagine physics as a great city and Mikola as a professional guide who knew only his own neighborhood. Still, Mikola's class was quite useful. It was the last formal physics course I ever took.

András Kubacska was a tall man who taught zoology and botany. I was glad when we reached botany in our fourth year. I liked the evergreens and gymnosperms. Kubacska knew less than Rátz or Mikola, but he knew the common plants and animals and made pleasant presentations. For years, I liked botany almost as much as mathematical physics.

Kubacska had a son in my class, and as a small favor to the family I tutored the boy in mathematics. Both of us struggled with this task; the boy lacked the ingenuity of a first-rate student. But we liked each other and corresponded for many years after.

The Eötvös Prize in Mathematics was an important prize

named for Loránd Eötvös, a physicist famous for his work on gravitation. Eötvös was then an old man, near death. The Eötvös Prize was a kind of Hungarian national mathematics prize given at the close of high school.

But you had to compete for it, and I did not compete, either for the Eötvös Prize or any other prize. I felt instinctively that prize seeking was conceited. If you are worthy, the prize will seek you.

My description of the Lutheran gimnázium should include a fuller portrait of Jancsi von Neumann. Because Jancsi must have been far more outstanding in gimnázium than Edward Teller, Leo Szilard, or anyone else I have ever known. He was truly a miracle.

When I met Jancsi, I was about 13 and he was about 12. He was one grade below me, but in mathematics two classes ahead. He already had an astonishing grasp of advanced mathematics. Budapest had a strong mathematics community, and Jancsi made himself well known in that circle before he even left gimnázium.

I never felt I knew von Neumann well at gimnázium. Perhaps no one did; he always kept a bit apart. He loved his mother and confided in her, but hardly in others. His brothers greatly admired him, but they were not intimates. Jancsi's physical coordination was poor, so he did not make friends through athletics. He joined in class pranks just enough to avoid unpopularity.

Jancsi loved money and material things. He was glad to have a banker for a father. Besides money, his interests were almost entirely intellectual.

I walked home from school with Jancsi about once a month. The depth of his mathematical knowledge amazed me. The way he described set theory and number theory was enchanting. The beauty of the subject, his intensity and facility of description made me feel we were close friends.

Jancsi went on and on and I drank it in, hardly speaking myself. Both his knowledge and his desire to relate it seemed inexhaustible. Most people walk straight home, already thinking of what they will do when they arrive. Not Jancsi. One got home late after a walk with him.

I have known a great many intelligent people in my life. I knew Max Planck, Max von Laue, and Werner Heisenberg. Paul Dirac was my brother-in-law; Leo Szilard and Edward Teller have been among my closest friends; and Albert Einstein was a good friend, too. And I have known many of the brightest younger scientists. But none of them had a mind as quick and acute as Jancsi von Neumann. I have often remarked this in the presence of those men, and no one ever disputed me.

You saw immediately the quickness and power of von Neumann's mind. He understood mathematical problems not only in their initial aspect, but in their full complexity. Swiftly, effortlessly, he delved deeply into the details of the most complex scientific problem. He retained it all. His mind seemed a perfect instrument, with gears machined to mesh accurately to one thousandth of an inch.

Despite his singular achievements in mathematics, Jancsi was raised to be well rounded. He knew English almost as well as Hungarian, perhaps from a tutor engaged at home. Eventually, he also spoke fluent German, French, and Italian.

But Jancsi's intelligence never appalled me. He was clearly better in mathematics than I was. But so were many others; I was not a mathematics prodigy. And I knew more physics than he did. I did not compete with von Neumann for prizes, scholarships, or positions. If we tried, in a friendly way, to persuade each other of certain things, that is not competition.

*       *       *

Almost as soon as I graduated from the Lutheran gimnázium in 1920, I felt a terrible letdown. I had lost daily contact with mathematics, physics, and most of my friends. I told my father that I needed more education. My father deeply respected learning and the benefits of an advanced degree. He agreed to help me continue my schooling. I enrolled for a year in the University of Technical Sciences in Budapest, called the Muegyetem.

I attended physics lectures, but like most Hungarian universities, the Muegyetem was not interested in modern physics. After two years under Sándor Mikola, the Muegyetem course seemed like a repetition. I wanted better.

I had begun to consider an occupation. Coming from a family recently Jewish, I could never be a Hungarian politician. But I had no wish to rule. Jews were also not allowed to be policemen or soldiers. But I had no desire for those jobs either. War makes a poor occupation. In peacetime, a soldier takes excursions, it is true. But his purpose is not to take excursions; his purpose is to kill other men. This did not attract me. The policeman's job looked nearly as bad: subduing rogues and criminals by force.

Well, what jobs were open to me? Certainly I might have had a fine job at the Mauthner Brothers tannery. But that would have been too familiar. To be a farmer might have been satisfying. But I had no land and would not have enjoyed taking orders from a landowner.

I could not have been a professional singer and did not want to be. There seemed only a few hundred actors, singers, or dancers in all of Budapest, and I knew none of them and had no talent at all in that direction.

I had no chance whatever to be a professional athlete. I was still somewhat sickly, and at 5 feet 6 and 120 pounds, I cut an unimpressive figure. I had never played football well. Already, many schoolboys could run and jump better than I could.

I lacked the talent to write novels like Mór Jókai. Reading his books, I knew that I could never produce anything similar. Nor could I write good poetry.

I could have been a lawyer. I had an uncle who was a lawyer. But what was the law but a tiresome quarrel with another lawyer? People claimed that the law touched great social issues, but the cases I saw only tried to squeeze a bit of money out of some argument or treaty. It seemed a quarrel job.

I could have been a doctor like Grandfather Einhorn. And I knew that curing the sick was noble work. But the diagnoses were too hard for me and the hours too long and uneven. I was glad that Grandfather Einhorn never urged me to become a physician.

Neither did I want to be a clergyman. I liked a good sermon. But religion tells people how to behave and that I could never do. Clergymen also had to assume and advocate the presence of God, and proofs of God's existence seemed to me quite unsatisfactory. People claimed that He had made our earth. Well, how had He made it? With an earth-making machine?

Someone once asked Saint Augustine, "What did the Lord do before he created the world?" And Saint Augustine is said to have answered, "He created Hell for people who ask such questions." A retort perhaps made in jest, but I knew of none better.

I saw that I could not know anything of God directly, that His presence was a matter of belief. I did not have that belief, and preaching without belief is repulsive. So I could not be a clergyman, however many people might gain salvation. And my parents never pressed the point.

I asked myself: "What am I principally interested in?" It was clearly physics and mathematics. So I decided to become a physicist. But in 1919, aspiring to be a physicist was considered a bit crazy. Why should an intelligent, well-educated man from an honest family make sacrifices to join a profession with few positions and dismal salaries?

A brother of one of the Mauthner Brothers tannery owners was a university teacher, one of the few scholars that my parents knew personally. This man was no genius, but capable. He had never been able to rise above the rank of privatdozent, a kind of unmarried assistant professor. His chief trouble was that he was Jewish; the university clique had stunted his career because of it.

Seeing this made my parents leery of their son embarking on an academic career. They could not see the great compensations of physics: the joy of the work, the pleasure of grasping the fundamental nature of things, of exploring and developing that understanding.

But I saw those things and felt them deeply. So when my father asked me, "What do you want to become, my son?" I said, "Father, if I am honest, I want to become a physicist." I had in mind not a teacher of physics but a scientist working on the theoretical and experimental frontier of physics.

My father seemed to expect this answer. After a minute,

he asked, "Well, Jenö, how many jobs for such physicists exist in our country?"

With a little exaggeration I said, "Four." There were really only three. Two in Budapest, and one in Szeged, at the universities.

Then my father said, "Jenö, do you think you will get one of those four jobs?"

How could I respond? I knew how badly I wanted one of those jobs, how hard I would work to get one. Perhaps I might succeed eventually. But how could I be sure? I was just 17 years old, near the top of my class in one small school. And already I knew one boy far more intelligent than I was: Jancsi von Neumann. Von Neumann had decisively shown me the difference between a superb mind and all the rest of us.

And at 17 I was still securely under the influence of my father. So I readily agreed to the idea that he pressed on me: That I would study not physics but chemical engineering and prepare to follow my father into the Mauthner Brothers leather tannery.

∽ *Five* ∽

# Albert Einstein Made Me Feel Needed

The Mauthner Brothers tannery had been founded around 1750 by Hungarians of German descent. It had kept a strong faith in German science and industrial culture ever since. My father had absorbed this attitude.

After a family vacation around 1920, my father brought me to Berlin, with an eye to my future. Together, we looked at several institutes of technology. We walked through the Charlottenburg district, just west of the old city. I recall the buildings there, sober and grand.

My father was visibly impressed with one institute called the Technische Hochschule, or School of Technology. Looking it over, he decided that his tannery partners were not wrong about German industrial virtue. My father also liked the idea of his son leaving Hungary for a time. He wanted me to explore another country. See the world a little, he said. He knew that I would learn more in Berlin than I could in any Hungarian technical institute.

And already he disliked the drift of Hungarian politics. It was a trying time for men who hated revolutions. I could have gone to Switzerland for schooling. Zurich had a fine hochschule, which Jancsi von Neumann once attended, but the Wigners thought Switzerland a bit too exotic and expensive a locale for a young student.

Germany had never known much immigration or been known for its warm embrace of foreigners. But in 1921, it seemed a land of some political resilience. It had a vocal communist group, but then each of their two uprisings had been handily suppressed by the German army. So my father made a mistake that many Jews made in the 1920s: He picked Germany as a safe haven from communism and revolution. And I did not dispute my father.

In 1921, at the age of 18, I moved the 950 kilometers from Budapest to Berlin to study chemical engineering at the Technische Hochschule. I never intended to live outside Europe, to help fight any war, or work for any government. My father still expected me to follow him into the Mauthner Brothers tannery. I hoped to become a respected physics professor at a German or Hungarian university.

My parents knew that physics was my true love. They knew, but they hardly approved. If my mother approved, it was only for my happiness. She had not finished high school; science was something that she did not know how to think about. My father hardly approved because his tannery did not hire physicists. But my parents never kept me from physics. And as I worked among physicists in Berlin, I grew more intent on becoming one myself.

In 1921, Berlin was not yet split into East Berlin and West Berlin. But it was already a fractured city, poorly connected by subway. Since 1871, it had been the capital of the German empire. Now it was the capital of the new Weimar Republic.

The streets of Berlin were crowded with strangers. I did not shrink from such crowds, but neither did I absorb from them anything practical. My rooms were private, and nearly all of my friends were students. I rented an apartment in a plain three-story building. Breakfast was made by the land-lady. Lunch and dinner I took in cheap restaurants—soup, meat, and vegetables, with a sweet for dessert. Drinking beer was often expected. The Germans liked to say:

"He who does not love wine, women, and song/
    Remains a fool all his life long."

I could not fully agree with this sentiment, but I admired it.

The Technische Hochschule gave a strong, practical training. I took a mathematics course in differential equations from a professor named Rothe. But chemistry was the main lecture and examination subject. My organic chemistry class was well taught by a chemist named Pschorr, of the Pschorr brewing family.

I learned chemical analysis and explored various laws of nature. In my apartment, I read books and articles on chemical analysis, set theory, and theoretical physics. Herz's book, *Physical Chemistry,* and K. A. Hofmann's new *Textbook of Inorganic Chemistry* I knew thoroughly.

I was part of a small student circle. One friend was a man named Popper, another a chemical engineering student, also Hungarian. Popper was an aspiring Jewish businessman who wanted a broad scientific education.

My father had sent me to learn the chemical engineering of leather tanning. I tried to learn it. It was never a devotion, but I mastered much of the tanning process, partly from reading books outside the curriculum.

The organic chemistry professor at the Technische Hochschule was famous as a chemist, and justly so. I heard a few of his lectures. But after a time, I hardly went to lectures at the school. German technical schools did not require lecture attendance so long as you passed your examinations.

Theoretical physics was what fascinated me. Engineering I saw chiefly as an application of physics. Studying mostly physics was rare at the hochschule, but it was accepted. A few other young men did the same.

The quality of my work was not remarked on. Rátz, at the Lutheran gimnázium, had never called me a prodigy. Neither did my teachers at the Technische Hochschule. Why should they? I was a quite ordinary young man who did just what was expected.

I studied chemical analysis in my inorganic chemistry course. Inorganic chemistry was a tangible discipline then, not the mass of electronic orbits that it has become. The inorganic chemistry course had three parts: a wonderful text, a lecture, and, most important, a laboratory.

Laboratory work was a consuming duty. Six days a week for two years I reached the chemical laboratory in the early morning. The room had five tables. I worked at the same end table, finding the chemical constituents of a small compound or perhaps analyzing the gaseous compound sulfur hydrate. We got used to its unpleasant odor.

I worked there until midday, ate luncheon there, and went home after five o'clock quite tired. I had many analyses to complete before I could qualify for my examinations. A laboratory instructor was present to resolve problems; but then you were not expected to have problems. The wonderful text clearly presented the material; the lectures ably supported the text. I worked hard in the laboratory and did fairly well. But chemical analysis always felt like a duty. The instructor on my final oral examination knew how much I liked the crystal structure of sulfur, and obliged me by asking about the chemistry of sulfur. I passed the exam.

One physics examination I recall distinctly. Students were then expected to stand at attention when a teacher entered the room. Four students were taking this exam; I was the first to arrive. Each time another student entered the room, I jumped to my feet, expecting the teacher. That is the kind of young man I was.

I tried to read more widely than a young engineering student might. The makeup of human societies interested me, and I thought vaguely of conducting research in anthropology. One book that terribly impressed me was Sigmund Freud's *The Interpretation of Dreams.* Freud gave the answers to questions that a dull young man like Wigner had never even considered. Freudian psychology is an artful creation which I have admired all my life. How many works of art have been so deeply useful? And I admired the great care with which Freud observed his patients. He was not only a genius but a diligent man.

Freud's work led me to think deeply for the first time about human consciousness. Consciousness is that thin layer of experience no greater in ourselves than is our small planet in the mighty universe. Yet when we speak of ourselves, we refer almost exclusively to this thin layer. I have never lost my fascination with human consciousness.

I balked at some of Freud's conclusions. That dreams express and then seek to satisfy our desires was a great notion. I felt that Freud undervalued the virtues of an untroubled sleep. And Freud debased love by linking it too strongly with sexual desire and financial advantage. Still, Sigmund Freud was clearly a genius. Alone, he had founded a new science— and how many men have ever done that?

Reading my own dreams proved fruitless. Those I recalled were simple life reflections. But I agreed with Freud about the power of the unconscious mind. For years I had found intractable problems could often be solved by rest and a healthy walk. As I struggled to learn physics, I noticed that ideas taken from the half-conscious minds of physicists often proved decisive.

Sigmund Freud's work was so inspiring that I thought vaguely of becoming a psychiatrist, or at least a theorist of the mind, studying consciousness. I might have done so had I not become so engrossed in the curious world of physics.

⚜ ⚜ ⚜

I had little free time as a student because I wanted to hear other lectures. One event I took special care to attend was the Wednesday afternoon colloquium of the German Physical Society, held at the University of Berlin. This university was located quite near the Technische Hochschule physically, but they never shared an intellectual life and their professors did not mingle. The Technische Hochschule was devoted to applied science; the University of Berlin to pure science.

The physics colloquia of the German Physical Society—what a departure they were from the routine of the hochschule! About 60 people came to these colloquia. Few of them knew all the others. Some days it was too crowded to find a seat.

At my first colloquium, I hardly understood a word. I kept hearing the buzz of strange phrases like "ionization energy." Yet somehow I was fascinated. I felt that I belonged there. Not yet to speak, but to listen and observe.

I can still see the meeting room as it was that first day: a sizable classroom largely filled with three sets of wooden chairs; the chairs in each set were connected. I sat, listening, in a middle row.

This was my first view of Albert Einstein. Seated next to him was Max von Laue, a man now far less famous than Einstein, but in 1921 nearly as well known, and deservedly so. Von Laue was 42 years old that year. He had won the Nobel prize in physics in 1914 for discovering the diffraction of x rays by crystals. Von Laue had taught in Munich and Zurich as well as in Berlin.

And so many other great scientists came to the colloquia: Max Planck was one, a man in his 60s, the permanent secretary of the Prussian Academy of Sciences and the Nobel laureate of 1918. Einstein had great respect for Planck, and von Laue had gotten his doctorate under Planck back in 1903.

The physical chemist Walther Nernst often attended. Nernst was a plump little balding man with a ready laugh and a bushy mustache. Back in 1893, he had published a famous little book called *Theoretical Chemistry,* which grew fatter each time he updated it. Nernst had just been awarded the Nobel prize in chemistry for 1920.

Rudolf Ladenburg, who became a good friend of mine years later, was often there. So was another future mentor, Richard Becker. Werner Heisenberg came when he was in Berlin, though I was not yet familiar with his work. Wolfgang Pauli came also.

Pauli was a young man from Vienna, just a few years older than I was, but treated as a prodigy. By 1923, he would be a lecturer at the University of Hamburg. "Don't be too proud," Pauli used to warn those who theorized boldly at the colloquia. "What you say doesn't work." Pauli was derisive, but his words stirred discussion, and beneath the derision he had a good heart.

Moving easily at the head of it all was Albert Einstein. He

had been working in Berlin since 1914. In the years since, scientific positions all over the world had been offered to him, yet he chose to remain in Berlin.

All the rest of us were in Einstein's shadow. Just knowing him well helped you to find a job in physics. Most great men are respected, but Einstein also inspired real affection. He had a great many lovable traits.

At gimnázium in Budapest, I had been told that atoms and molecules were irrelevant to physics. My physics course there had nearly ignored the great contemporary physicists. We had learned of Isaac Newton and how he calculated the velocity and future position of the planets. But we had learned very little of men like Einstein and von Laue.

But in Berlin, I not only learned of the great contemporary physicists, I saw them every week with my own eyes, heard them speak with my own ears, and studied some of the same questions they did. These were physicists who not only believed in atoms and molecules, but found them important and were sure they could be observed. All this was quite invigorating.

I began reading a great deal of new material and soon enough I could follow most of the colloquia discussion. I even helped formulate several papers. One of these papers, on the equilibrium of a piston with black-body radiation, was published by Wolfgang Pauli. My contribution was not at the heart of the paper. Still, I thought, "Well, I am not so stupid after all."

Max von Laue ran the colloquia. Von Laue read the titles of four or five important new physics papers and asked a different man to read each paper and to prepare an oral review of it for the following Thursday. The reviewers had to understand the papers and be able to convey their spirit.

Von Laue himself once prepared a review on something

we called the Einstein–Ehrenfest paradox. And Einstein once reported on the so-called Bose statistics, making a small joke as he talked. But most of the reviewers were junior physicists. These reviews allowed the major physicists, who were absorbed in their own work, to still follow other branches of physics.

One Thursday, I was chosen to prepare one of these reviews. I got a bit excited over that. But looking back at those days, I am surprised that I was not more nervous to be explaining physics to Albert Einstein and Max von Laue. I was too young to see the historic meaning of these colloquia.

In the first row of chairs sat Einstein. Next to him were Nernst, Planck, and von Laue. Einstein was a man of average size, just over 40, with a pleasant, open face and a quiet, attentive manner. But for his unruly head of hair, he looked quite normal.

If the reviewer presented a clear picture, no comment came from the first row. But if the review of the paper was unclear, questions were sure to arise from the first row, especially from Einstein. He was always ready to comment, to argue, or to question any paper that was not impressively clear: "Oh, no. Things are not so simple." That was a favorite phrase of Einstein's.

But Einstein's clarity of thought and skill in exposition were matched with a simplicity and an innate modesty. He could have made a great show of his own importance. He never thought to do so. He did not want to intimidate anyone. On the contrary, he accepted the logic of a colloquium: that human intelligence is limited; that no man can find everything alone; that we all contribute. Perhaps that is why I never felt nervous at the colloquium. Albert Einstein made me feel I was needed.

Einstein's modesty was being tested in 1921 by his great

and growing fame. He had already recast the very foundations of modern physics. He was awarded the Nobel prize that year for finding the photoelectric effect; and yet we knew that the photoelectric effect, inspiring as it was, was not his masterwork.

In 1905, Einstein had first become famous with the Special Theory of Relativity. Now, others had defined pieces of the theory before. Nearly all physical ideas have antecedents. But no one before Einstein had seen the breadth and stature of relativity. Einstein saw this at once and richly expanded and sharpened the theory. So he deserves the title of founder of the Special Theory of Relativity.

Around 1915, Einstein had given us the General Theory of Relativity. Like many of his equations, it needed some adjusting. But Einstein had seen the essential fact: Simultaneity is not absolute. Events appearing simultaneous to one person may not be simultaneous to a second person in motion with respect to the first. Einstein had stated this idea with emphatic clarity and it had deeply stirred our little world.

Every physicist read Einstein's work, *Modifications in Relativity Theory.* And in the ten years between the Special and General Theories of Relativity, Einstein had published more than 30 papers on other subjects: statistical mechanics, quantum theory of radiation, solid-state physics, opalescence, and electrodynamics. And though he had not yet attempted an all-embracing basis of theoretical physics, his interest in that notion was clear.

New ideas are rarely fundamental, but Albert Einstein conceived ideas that were both fundamental and new. A moving body has the same physics as a body at rest: This is a terribly simple idea. Yet it took deep insight to see it.

Einstein was almost impossible to surprise in the realm of

physics. He seemed to foresee everything of major importance and to describe it with fresh, startling work.

Most people do not grasp great concepts all at once. They see only a small bit at a time. The core of an idea appears to them in a moment of inspiration, but it is months or years before they have polished their work, explored its immediate implications, resolved its most evident flaws.

Einstein did not work this way at all. Concepts seemed to occur to him, fully realized. Their flaws and implications he saw immediately. He, too, worked to polish his work, but playfully, with a clear idea of what he would find in the end. So Einstein was famous in 1921, and becoming more so.

Fame is a complicated idea; and being more than mildly famous is highly unnatural. A famous man suffers a kind of permanent intrusion. The public holds grossly simplified images not only of his work, but of his personality. Many in the public half-consciously regard him as a personal friend and approach his personal affairs with a possessive concern.

Albert Einstein's friendliness had indirectly encouraged some of these phenomena. People who knew almost nothing about physics were drawn to Einstein's human image. So Einstein knew very well that some of his fame was hardly deserved.

He also knew that his life was well timed; that 5000 years before, men had lived with genes like his. They had not left us theories of relativity because the discipline of science had not existed then. Einstein knew that he was lucky to enter a world that honored science and had already given it a foundation. All these realizations made Einstein want to promote his ideas and not his fame.

Before I attended the colloquia of the German Physical Society, I had thought that the greatest physics was mainly experimental. The Nobel Committee still stressed experimental work in those days. And the physics books I had read before attending the colloquia had treated quantum theory with some condescension. They raised the age-old prejudice against theory: That it is important "only in theory," not in practice.

But I noticed that the great physicists at the Thursday colloquia were mostly theoretical. They spoke of quantum theory often and with great respect. They seemed to regard theory more highly than practice.

Their attitude opened my eyes. Already I knew that I lacked the patience to ever be a great experimentalist. Defining experiments requires theory, but the primary need of the experimentalist is patience. The work is slow. You hope that an unusual result will prove some new theory, but you know very well that most such results come instead from a flaw in the testing procedure. I never had the patience for that. I wrote my theoretical papers quickly. I had an idea and quickly set it down on paper.

To a young man of my temperament, theoretical work was far more promising. I saw that I might someday have the wisdom and imagination to excel there. Not to be an Einstein or a Heisenberg, but to gain respect as Jenö Wigner. I would never have been noticed as an experimentalist.

Colloquia discussions were lively, with something of a dilettante spirit. We wondered how to reconcile light interference and the photoelectric effect. We discussed the ionization energy of mercury, the second excited state of thallium, the Bohr–Stoner periodic table, and the theory of the hydrogen molecule. And we spoke of people now largely forgotten: Hund, Witmer, Dorgelo, names only dimly familiar to physicists today. We knew them well 65 years ago.

One element missing from the colloquia was concrete

encouragement. Einstein was very kind to young physicists, but even he did not push us along as he might have done. He never said, "Look here, this idea of yours is quite promising. Why don't you work it out and publish it?" I waited in vain to hear such words.

And Max von Laue was a bit the same way. He accepted thesis students, but he made clear that he felt physics was in a sad state and he no longer assigned thesis topics. So, very few people formally studied under von Laue. I never knew Max Planck to have a student and Walther Nernst did not like teaching either.

Einstein, Heisenberg, and many of the others were as brilliant and subtle as any diplomat or statesman. Yet it was a group that avoided politics and worldly affairs. Science was a monastic occupation then, with a monastic spirit. Its practical benefits were obscure. People thought we were a bit crazy to devote our lives to physics. We knew our reputation, but what could we do? We loved science and felt compelled to devote ourselves to its advance.

Colloquia were then commonplace. Most scientists worked longer hours than scientists do today, but we still found time to form small scientific societies that spread a love and knowledge of science. Max Volmer, one of the great physical chemists of the 1920s, had his own colloquium then, for which I reviewed at least one paper.

Even when the colloquium was over, we did not scatter. Many of us went out to a coffeehouse and sat around a large table, talking further. Conversation was not always rigorous, but it touched on all of the things we loved, not only physics but nature, family, and culture.

My third year at the Technische Hochschule, I arranged to work about 18 hours a week in the physics building at the Kaiser Wilhelm Institute. The institute was in the Berlin suburb of Dahlem. It was a kind of German precursor to the Institute for Advanced Study in America, with sturdy buildings in a number of different fields: chemistry, physics, and so on.

And there at the Kaiser Wilhelm Institute worked a man who decisively marked my life: Dr. Michael Polanyi. Few people in this century have done such fine work in as many fields as Polanyi. After László Rátz of the Lutheran gimnázium, Polanyi was my dearest teacher. And he taught me even more than Rátz could, because my mind was far more mature. After Rátz and my parents, Polanyi was my greatest influence as a young man.

The Germans have a tremendous word for fiber chemistry: "faserstoffchemie." Michael Polanyi had his own laboratory in the Kaiser Wilhelm Institute for faserstoffchemie. The Mauthner Brothers tannery in Budapest employed a fine chemical engineer named Paul Beer, who somehow knew Polanyi and gave me a strong letter of introduction to him.

So Dr. Polanyi asked me over to his home one evening. A chemist named Herman Mark also came that night. Mark was an energetic, chatty man from Vienna. He was only seven years my senior, but seemed much older.

Mark had fought in the Austrian ski troops during the First World War on both the Russian and Italian fronts and had escaped from an Italian prison camp disguised as an Englishman. He had quickly completed his education at the University of Vienna and taught at the University of Berlin before joining the Kaiser Wilhelm Institute as a research associate.

Polanyi and Mark had a fabulous discussion that evening, just two physical chemists discussing one topic after another. Mark smoked a few cigarettes. I sat by without opening my mouth, amazed at how much physical chemistry they knew. Topics at the farthest edge of my comprehension they discussed with the greatest fluency and ease. They spoke with a graceful insightful wit, following each other perfectly.

When Herman Mark finally rose to leave, my involuntary reaction betrayed my great disappointment. Mark put on a little half-smile, sat down again, and revived the conversation. My embarrassment at having kept Mark in the room soon faded in the face of their startling conversation. Listening with all of my limited intelligence, I knew that I was deeply happy.

That was my introduction to Dr. Mark and Dr. Polanyi. Soon I knew Polanyi closely. He told me to call him "Misi" (pronounced "*Mee*-she"), placed me in his laboratory, and asked me to contribute to meetings and colloquia.

About three other students worked for Polanyi. I studied theory: crystal symmetries and the theory of the rates of chemical reaction. I spent just a few hours in the lab and many more hours calculating figures in my room. I also learned a great deal about the life of Michael Polanyi.

Polanyi had been born into a Jewish family in Budapest in 1891, 11 years before me. His father was a civil engineer and traveling railway entrepreneur, who had brought home to his children vivid tales of German science and industry. But Mr. Polanyi had abruptly lost his fortune in 1899 and died in 1905.

So Michael Polanyi had spent the last years of his boyhood with very little money, but surrounded by poets, painters, and scholars. He had graduated from gimnázium in 1909, entered the University of Budapest as a medical student, and found that he preferred biology and physical chemistry to medicine. So Polanyi had studied chemistry in Karlsruhe, Germany.

While I was watching the First World War from the Lutheran gimnázium in Budapest, Polanyi was serving in the war as a medical officer. After the war, he had taken a chemistry doctorate at the University of Budapest.

But the communist rise in Hungary, with its threat of violence, sent Polanyi back to Karlsruhe in 1919. By then, he was an authority on the adsorption of gases. He had joined the Kaiser Wilhelm Institute in 1920, married, and soon settled in the Kaiser Wilhelm Institute for Physical Chemistry and Electrochemistry.

Polanyi and I wrote a joint article in 1925, introducing assumptions that seemed drastic then; they later proved quite correct. We wrote another joint paper in 1928. What a pleasure it was to assist a man of such keen mind and deep insight.

Polanyi took an interest in all of his assistants, but I felt that he liked me especially. He freely advised me on various personal matters. In time his generous wife did too. Polanyi even loaned me a bit of money when I needed it.

But his finest gift was to encourage my work in physics, and this he did with all of his very great heart. In all my life, I have never known anyone who used encouragement as skillfully as Polanyi. He was truly an artist of praise. And this praise was vital to me because it was often missing at the great afternoon physics colloquia.

Because Polanyi was a decade my senior and held a far higher position, it was not quite proper for him to befriend me as he did. But Polanyi cared nothing for formal questions of age and status. That was part of his great sweetness. Polanyi was concerned instead that young men should love science and labor to understand it. He was concerned that he could never fully share his love and the knowledge he had gathered.

Like me, Polanyi enjoyed asking questions outside the realm of basic science: Why is the world divided into separate nations? Why do all nations have governments? How should a man live his life in a world filled with evil? Polanyi even taught me some poetry. He made learning a great pleasure.

Dr. Polanyi and I did not always see eye to eye. Polanyi found quantum theory too mathematical for his liking. I was the only one in his lab deeply interested in it.

Once I made an observation to Polanyi about the impossibility of an association reaction. He heard my idea without grasping it. I felt sure that I was right and even that my idea had merit. But I was too modest to press it home.

Months later, Polanyi told me one day, "I am quite sorry. This point which you have always made on association reactions: I have just heard it in a paper of Born and Franck. I told them that you had the same idea, but they have already sent in the article, and nothing can be done." Polanyi paused a moment. "I am quite sorry," he said again, "I don't know why I failed to understand you."

Well, I think I know. Even a man as open-hearted as Polanyi does not easily accept the brash ideas of a modest and untried assistant. What I had told him was radically new, and however open-minded people may seem, very few are prepared to embrace radical ideas.

I hardly minded losing credit. I kept studying the problem, glad to have men like Max Born and James Franck as colleagues of a sort. Born and Franck were just great names and distant images to me then. Franck was awarded the Nobel prize in 1925; Max Born later earned one, too.

I worked up a slight variation on this idea in my doctoral dissertation. Max Volmer, reviewing the dissertation, took my claim skeptically. He believed in association reactions and he marked this part of the thesis: "Needs better foundation."

I published my results anyway, hoping that if I could not impress Max Volmer, I might still impress Max Born and

James Franck. I was puzzled to find them little impressed. They responded: "Maybe so, maybe not so."

Polanyi advised my doctoral dissertation at the hochschule. I chose a topic far from the crystallography of Weissenberg or Herman Mark: chemical reaction rates. I wondered: How do colliding atoms form molecules? We knew that hydrogen and oxygen make water in a container, but how soon? How much depends on pressure and how much on temperature? I pursued such questions with elements far more complex than hydrogen and oxygen.

Polanyi was a wonderful advisor. He understood chemical reaction rates both in theory and practice. He accepted my proposal that angular momentum is quantized and that the atoms collide in a proportion consistent with Planck's constant. This idea is now widely known, but then it was rather brash. And studying chemical reaction rates taught me much about nuclear reaction rates that would be useful in future years.

My thesis paper for the engineering doctorate was submitted, with Polanyi's name attached, in June 1925. We called it "Bildung und Zerfall von Molekülen" ("Formation and Decay of Molecules").

The engineering degree was called a "diplomarbeit." I was able to take my diplomarbeit by helping Herman Mark study the lattice structure of rhombic sulfur crystals.

I came to know Mark better. He was a pleasantly vigorous man who liked food and wine, games, and song. Mark's father, like my own, was a Jew who had converted to the Lutheran faith. Herman Mark had a bit of the Hungarian kings in him. Like a king, he was scrupulously honest, ready to act, and fond of those in his care; but again like a king, he saw no need to consult others closely or to justify his actions. Mark knew at least one thing better than Michael Polanyi: that a teacher should not overly indulge his students. Mark patiently refused the brasher requests of his pupils.

Mark chose not to closely supervise my thesis. We talked generally about my topic. Crystallography, the science of crystal structure, is wonderfully full of symmetry, and I told Mark how fascinated I was by the crystal structure of sulfur. My thesis was unusual for being microscopic. And the way that sulfur atoms arrange themselves was then nearly unknown; naturally, I was pleased to advance the field, even slightly. Mark assured me that he was quite pleased to see me so fascinated with my work, but I must pardon him if he was somewhat less fascinated with it. Mark was not an easy man to impress. He had seen similar work before and was awfully absorbed in work of his own.

So Herman Mark was a strong teacher, but Michael Polanyi was really the miraculous one. Polanyi loved to ask the fundamental question: "Where does science begin?" He listened to the thoughts of others on this question, but he also had his own well-crafted answer: "When a body of phenomena shows coherence and regularity."

Polanyi loved and honored the scientific method with great truth and devotion. He managed to keep all of science within his fond gaze and a great deal more besides. What a mentor Michael Polanyi was.

# Six

# *Learning from Einstein*

$\mathcal{G}$reat events surround us in our youth and we cannot follow them all. We lead our lives, largely ignorant of what will intrigue future historians. Some of the history that surrounds us we will never grasp; much of the rest we grasp only vaguely, and then only years later.

Historians tell us that Berlin in the 1920s was a city in chaos. Perhaps it was. Perhaps the workers were discontent; perhaps the rulers were nostalgic for past glory; perhaps the middle classes yearned for a return to that golden age when Germany was the leading power on the continent.

If so, I was largely unaware of it. I am not an outgoing man and was even less so as a youth. I knew neither the rulers nor the workers of Berlin. I knew awfully few Berliners at all. Most of my friends were students and many of them were foreigners. I ought to have ventured beyond this small group, but I felt too busy and shy to do so.

I do recall the terrible inflation. In 1923, the German

mark fell to pieces. There were 4500 marks to the dollar in late 1922 and more than 4 *trillion* marks to the dollar just over a year later. The Germans resented this inflation deeply, and the disruptions it caused in their national life were clear even to a young man like Wigner.

But the years 1924–1929 were more stable in Germany. Inflation was brought under some control. Germany rejoined the League of Nations. To have seen deep trouble in the social and political currents of Germany after 1923 required a man more subtly discerning than I was.

I considered myself a visitor in Germany, a student of physics, not politics or economics. Most of my money came in foreign currencies. My stipend from the Physics Department was in German marks, but my father sent me Hungarian money. I even had some English money, which I exchanged at increasingly handsome rates as the German currency sank to nothing.

❧　　　❧　　　❧

Is a radical a man who repudiates the society of his parents and teachers? If so, then I was no radical in Berlin. I admired my teachers more with each passing year. I loved my parents and wanted to help them. To dream of pursuing a career that they had not chosen was a radical enough path for a youth of my background. I had no wish to be more radical than that.

But if a radical is someone who regards a traditional subject in a revolutionary way, then perhaps I was a radical, because quantum mechanics had transformed physics and I embraced quantum mechanics fervently.

Physics in the early 1920s was full of vexing contradictions, a very long way from the string of beautifully straight principles sought by Einstein. All of physics seemed a primitive confusion. In the fields of light interference and the photoelectric effect, "the facts" seemed contradictory, and many other fields held their own stubborn mysteries. We had insights into these mysteries, but no solutions.

All these years later, it is hard to recall the primitive state of physics then, just as it is hard to recall that man's chief purpose was once hunting wild beasts for his dinner. We know these things but we forget them because so much has happened since. Human societies are too often embarrassed by the modesty of their origins.

Fundamental change comes slowly. The art of photography came slowly. The printing press came slowly. But in just a few years around 1925, I saw physics revolutionized by quantum mechanics.

Even before 1925, the quantum theory of matter was in serious trouble. It could not describe the atomic structure of anything but hydrogen, and even that description was inadequate. We all hoped for a better theory.

The classical mechanics of Isaac Newton was brilliant physics, but it could not explain much that we knew about the behavior of electrons and nuclei within atoms and molecules. A new kind of mathematical physics was needed. A quantum is a unit of physical property—a bundle of energy, for example. So we called the new physics "quantum mechanics."

The discovery of quantum mechanics was a nearly total surprise. It described the physical world in a way that was fundamentally new. It seemed to many of us a miracle as great as any physics equation of the past.

Quantum mechanics had no one single inventor. The idea of quanta came from a general statement by Max Planck. Planck had announced an insightful theory of black-body radiation in 1900. In 1913, Niels Bohr advanced his quantum theory of spectra. Then in July 1925, Werner Heisenberg made a breakthrough. He proposed replacing the classical picture of physics with a radical theory of quantum mechanics.

Heisenberg was just one year my senior but I had never met him. At age 21, studying under Niels Bohr, Heisenberg had wondered: Why do spectral lines in a magnetic field split into groups of polarized lines? Many able physicists before Heisenberg had already wondered this. But no one had properly explained the phenomenon. Even at age 21, Heisenberg was not like other physicists. He introduced the concept of half-quantum numbers to explain the mystery.

At a scientific congress at the University of Göttingen, Heisenberg had been pressed to defend his idea. Many older physicists were skeptical, including Heisenberg's beloved teacher, Niels Bohr, who was a strong backer of his students.

Bohr was the greatest scientist in Denmark; he was awarded the Nobel prize in physics for 1922. He was also a devoted expert of the hydrogen atom and of applying quantum theory to spectral lines. But Heisenberg defended his novel concept of half-quantum numbers with great assurance and equal modesty. The concept of half-quantum numbers made his reputation.

Heisenberg had none of the haughty dignity of many great men. Rather than crudely modify familiar concepts, Heisenberg liked to describe quantum phenomena with his own distinctive creations. Speculation by Werner Heisenberg was always inspiring, even when it did not hold water. He treated the plainest elements of physics as objects of rare beauty.

Heisenberg's initial view of quantum mechanics applied only to observable quantities: energy levels and transition probabilities between energy levels caused by the absorption or emission of light. But Heisenberg's idea was soon eagerly taken up. Within a few months, Max Born and Pascual Jordan had enriched its detail.

Heisenberg's theory of observations is still crucial to theoretical physics. Like many great physical ideas, it touches upon philosophy, asking implicitly: What is the state of a microscopic system? And how much of it can a human scientist observe?

The state of a macroscopic system is obvious. A chair is macroscopic. It can be described as having four legs and so on. But what about a microscopic system? Quantum mechanics told us that the atom can also be described by the wave function. How then is that verified? What are the properties? How do they follow from this wave function? These were questions that deeply intrigued Werner Heisenberg. His answers taught many of us a great deal.

Heisenberg knew too much quantum mechanics to share his whole vision. His deepest understanding stayed within. But he taught us that the nature of quantum mechanics is fundamentally statistical; that we cannot track at once both the speed and position of an electron. This notion is at the heart of his famous Uncertainty Principle.

When Heisenberg published his article on the Uncertainty Principle, I saw almost at once that he had ended the quantum troubles. I picked up the telephone and called someone—it was either von Neumann or Leo Szilard. I said, "Now we can go to sleep. The problem is solved." I was sure that Heisenberg's paper was exactly right about the behavior of electrons in the photoelectric effect; about the way that angular momentum is transferred to the constituents of an atomic beam; and about quite a lot more.

Fifty years later, the excitement I felt on first reading that article was still vivid. I cannot say just why the article was such a revelation. To blurt into the telephone that one man and one article have ended the quantum troubles is a massive statement. But I truly believed that.

I wish that everyone could have at least one such experience in this world: the dramatic, unexpected resolution of a nagging problem in their mental life.

In December 1925 and March 1926, the English physicist Paul Dirac proposed a slightly different quantum mechanics, emphasizing commutation relations. Dirac later helped create quantum electrodynamics and gave us many things, including the clever basic equation of the electron that we now call the "Dirac Equation."

Paul Dirac also became one of my dearest friends. He was quite tall, quiet, and modest. He moved slowly. But he knew his own excellence in physics. Dirac rarely discussed physics with anyone. It is very hard to do first-rate scientific work that way, but Dirac managed. He made very few mistakes.

Not every physicist loved Dirac's work. Some searched it in vain for a mathematical basis. Others complained that it lacked the plain, handsome rigor of fine scientific writing. One struggled to follow Dirac's work. He wrote his papers in English, while most of his readers preferred German. His papers used few English words, but to someone who barely speaks English, even a few dozen English words are irksome in a single paper on a complex topic. The language barrier, combined with Dirac's novel viewpoint, made his papers hard reading.

So at first Dirac's work was less admired than Heisenberg's. People said, "You know, there is a queer young Englishman who resolves these things in his own language. He may even be a genius. But what has he to offer to the honest German speaker?" Such was the common attitude toward Paul Dirac. Some people even whispered that he evaded the hard work of physics with clever tricks. What a foolish view that was! Dirac's work was indeed clever, but also rigorously beautiful.

Dirac had been inspired by Heisenberg. So I asked Dirac what he thought of that great Heisenberg paper, the solver of the quantum troubles. If Dirac had been conceited, he might have said that all of Heisenberg's thought was contained in the work of Paul Dirac. But he did not say that. Dirac said, "I think Heisenberg's paper contained a new physical idea." That was extremely high praise from a man as restrained as Dirac.

Dirac had used his own mathematical calculus to build a nearly entire system of quantum mechanics. His work stood as a splendid critique of Heisenberg's. But Dirac was an Englishman, who spoke German with an accent and corrected Heisenberg shyly. I think Heisenberg was slow to take Dirac seriously. Once he did, he saw how much quantum mechanics owed to Dr. Dirac.

Paul Dirac may have seen farther into quantum mechanics than anyone. His work had a beautiful, organic consistency. His culminating article was his famous light theory, connecting for the first time the radiation process with the mechanics of the atom.

Though the quantum mechanical articles of Born, Heisenberg, and Jordan were more widely read, Dirac's work was known even to his critics, who often measured new efforts against it. And that may be the highest compliment among physicists.

The signal contribution to quantum mechanics in 1927–1928 again was Dirac's. In February 1927, he introduced quantum field theory, which advanced not only light absorption and emission, but also beta decay.

The third man to deeply shape quantum mechanics was Erwin Schrödinger. Schrödinger was born in Vienna in 1887.

He had enjoyed superb teachers: Arnold Sommerfeld at the University of Munich and Max Born at Göttingen. Still, no one but perhaps Schrödinger himself expected him to swiftly change quantum mechanics. He had hardly touched the theory before. But in articles published in the late spring of 1926, Schrödinger recast quantum mechanics, suggesting an infinitely greater variety of quantum mechanical states than had ever been considered before.

Schrödinger's equation overwhelmed us. He had described quantum mechanics concretely and with mathematical clarity. He was inspired by existing equations and notations, and his work in wave mechanics especially was well-supported by the work of Louis de Broglie. But Schrödinger's equation was far clearer, more lively and convincing than any of the others. Eventually, Schrödinger created a differential equation to understand the quantized states of atoms and molecules. He wrote a series of brilliant papers, traveling along that road.

Heisenberg, Schrödinger, Dirac—how similar their goal, but how different their working styles! Heisenberg—working closely with his collaborators; daring, not always right, but giving a bold bird's-eye view and sweeping us along with his enthusiasm. Schrödinger—more solitary, but drawing on the work of others with a fresh, startling artistic spirit. And then Dirac—apparently alone with his brilliant logic. The wise among us learned from all three.

You may wonder: How did Albert Einstein regard all this quantum mechanics? He did not much like it. Einstein resisted quantum mechanics and the statistical nature of Heisenberg's Uncertainty Principle. He objected to a world where microscopic bodies no longer hold definite positions and where outcomes are described as probabilities. At first Einstein hardly bothered to master the novel work of Heisenberg,

Schrödinger, and Dirac. I think Einstein had once nearly endorsed the statistical view himself, before deciding it was flawed. Having almost accepted it once, he now rejected it not only with his mind but with his heart.

Einstein plainly saw that the statistical view was a quite novel way of interpreting physical events; he realized, perhaps even before many of its backers, that accepting the statistical view implied a need to reexamine a great many things, including human volition and desire. Einstein did not want to reexamine all that. So he made light of the statistical view. "How about the sun," he would say. "Is that also a probability amplitude?"

By 1928, Einstein no longer dominated physics as he had seven years before. He had heart trouble that year, stayed in bed for months, and afterward saw fewer visitors. This distanced him further from new trends in physics.

To his credit, Einstein slowly accepted the primacy of quantum mechanics in modern physics. His grasp on modern quantum theory was never strong, but he felt that quantum mechanics might be joined to the rest of physics, and even made some serious attempts at this. But that does not mean he liked it.

Einstein once said, "I shall never believe that God plays dice with the world." And he meant that very deeply. Einstein did not believe in a God that guides the lives of nations. But he certainly believed in a God that oversees the physical laws of the universe.

Einstein found the work of Heisenberg, Schrödinger, and Dirac quite clever. But they were dice throwers, and he did not like to see dice throwing. "The Lord God is subtle," Einstein once said, "but malicious he is not."

Einstein was consoled by his belief that all physical theories are temporary. He hoped that quantum mechanics would soon be superseded by a better, deterministic theory. That is part of the beauty of science: the knowledge that a deeper theory is always there, just beyond our view.

Until 1925, most great physicists, including Einstein and Max Planck, had doubted that man could truly grasp the deepest implications of quantum theory. They really felt that man might be too stupid to properly describe quantum phenomena. The recent work of Niels Bohr and others had slightly changed the perception. But still the men at the weekly colloquium in Berlin wondered: "Is the human mind gifted enough to extend physics into the microscopic domain—to atoms, molecules, nuclei, and electrons?" Many of those great men doubted that it could.

So finding quantum mechanics was an enormous relief. Now we knew: Man *was* bright enough to create a physics of quantum phenomena. Even without a final description, we knew that some day man would win, that he would understand atomic phenomena.

Science is a long progression of theories approaching the truth. Naturally, early attempts at quantum mechanics were spotty and limited in scope. But even in its infant state, quantum mechanics awed us. Charles Darwin had claimed that human evolution occurs only to assure our survival. Yet quantum mechanics did not aid human survival. Most of the physicists who had watched quantum mechanics evolve considered it a small miracle.

Miracles are hard to accept. Abandoning the causal conception of nature for a statistical one threatened our basic ideas about nature and pushed pure physics toward the realm of metaphysics. But most younger physicists accepted this bargain and struggled to follow quantum mechanics as it grew.

It was in my second year of chemical engineering study in Berlin that I first met Leo Szilard. In time, Szilard made himself a quite important man. I hope that you have seen his name long before this. But when I first met Szilard he was rather obscure.

Every Thursday, I saw Leo Szilard at the physics colloquium at the University of Berlin. Soon we introduced ourselves and got acquainted. My first impression was of a vivid man about 5 feet 6 inches tall, a bit shorter even than I was. His face was a good, broad Hungarian face. His eyes were brown. His hair, like my own, was brown, poorly combed, and already receding from his forehead. A full head of hair is quite nice, but we survive without it.

Szilard was born in 1898, so he was four years my senior. He rarely mentioned his father, a civil engineer and builder; but he had the deepest affection for his mother. Szilard had served in the Austro-Hungarian army during the First World War, then come to Berlin.

He had enrolled in the Technische Hochschule to study electrical engineering, but by the time I met him he clearly preferred theoretical physics. Szilard used to visit the analytical chemistry laboratory at the hochschule, for he was then infatuated with physical chemistry. But he had not yet settled into a profession.

Both Szilard and I had Hungarian for a mother tongue, and speaking Hungarian freely evoked the sweeter days of my childhood. Szilard also spoke a fluent German. He called me "Jenö," and I called him "Szilard." I noticed that he spoke with a striking clarity and vigor.

Szilard was an unusual Hungarian. He did not much like

trading jokes or taking long walks with friends. He was too busy to tell long stories. You might see him for a moment at a colloquium, but then he was gone. Several days later, he appeared at your front door with several bold ideas and not quite enough patience. Leo Szilard was always in a hurry.

The difference between myself and Szilard was clear in the way that we treated the physics colloquia. I very rarely sought an audience with the great physicists after the colloquia ended. In a crowd of 60 people, it seemed to me unreasonable to ask Albert Einstein or Werner Heisenberg to pay attention to Jenö Wigner. I was afraid of imposing. Looking back, I suppose this fear of imposing was unfounded. But I felt it keenly at the time. I was content to be introduced to Einstein's thoughts.

But Einstein's thoughts alone did not content Szilard. With typical directness, he sought out Einstein himself, shook his hand, and introduced him to Leo Szilard. Szilard introduced himself to many of the other top physicists as well. Even in 1925, Szilard felt that he was already someone important; so, he reasoned, all scientists would benefit from his acquaintance. He did them a favor by presenting himself.

I was not so sure that for Albert Einstein the meeting of Leo Szilard was a great favor. But I could not help but admire Szilard's fluency with great men. We seek out in others the traits we lack ourselves. And as Szilard often spoke to Einstein and I was often standing with Szilard, I, too, was soon introduced to Einstein. That was a magical moment for a young physicist.

❈     ❈     ❈

Einstein's appointment at the University of Berlin came from the Prussian Academy of Sciences. They made him both a university professor and a director of the Kaiser Wilhelm Institute division for physics research. That is quite a long title and it was even longer in German.

But this division for physics research barely functioned, and Einstein had no formal teaching duties. He lectured at the University of Berlin, but seldom. I think he wanted to teach more often, not only to share his wisdom with others but to expose himself to strong young colleagues who might challenge some of his cherished notions about physics.

It was likely Szilard who asked Einstein to give a seminar on statistical mechanics. The request was a bit odd because Einstein so clearly disliked the statistical nature of quantum mechanics. But Szilard, with his gift for courting men of prominence, made the request very naturally and directly.

Szilard had worked fruitfully with Herman Mark and had been accepted as a thesis student by Max von Laue, who rarely took thesis students. If Szilard had seen the president of the United States at a meeting or the president of Soviet Russia, he would have promptly introduced himself and begun asking pointed questions. That was Szilard's way.

Most physicists are shy and retiring, and this was even more true in the 1920s when physics was a far more obscure discipline. So Szilard's behavior stood out. People called him brash, but I think that the label "brash" misses his essence. A brash person cannot bear to be brushed off; but Szilard knew very well that sometimes a famous man must say, "You will excuse me. I am busy now." So I think that Szilard was not brash; a better word would be "well-relaxed" or "unencumbered."

Einstein agreed to teach the seminar on statistical mechanics in the winter of 1921–1922 and Szilard invited me to join. I was enchanted by Wassmuth's book on the subject. Several times in later years I tried to buy the book, but found it out of print. In Germany, many years later, I came upon Wassmuth's book. Rereading it, I found it rather mediocre, and I thought how sad it is to outgrow the influences of one's youth. But in 1921, I loved that book by Wassmuth. I absorbed myself in it in preparation for the seminar.

Einstein's decision to give this statistical mechanics seminar is one example of his great modesty. He could very easily have said, "No! My seminar will concern relativity theory, which is far more important." But as the reigning king of relativity theory, Einstein would have dominated the seminar. He would have lost the feeling of equality that he wanted. So he avoided relativity theory, and Szilard and I had the privilege of learning statistical mechanics from Albert Einstein.

What a splendid seminar it was! Einstein beautifully projected the spirit of the theory and showed us its inner workings. Then he pointed us toward the finest unsolved problems in the field, explaining not only their formal mathematics, but all of the related elements they seemed to hold.

The seminar students took turns presenting certain aspects of statistical mechanics. Einstein supplemented and clarified our presentations. His mood was unfailingly generous. Einstein worked simply and helpfully, more like a friend than a teacher.

Einstein's thoughts often turned philosophical. He told us once, "Life is finite. Time is infinite. The probability that I am alive today is zero. In spite of this, I am now alive. Now, how is that?" None of his students had an answer. After a pause, Einstein said, "Well, after the fact, one should not ask for probabilities."

We all knew that Einstein was a solitary man who liked to meditate on the world while walking alone. Yet I was one of many students whom he encouraged to address him personally and visit him at home. There we discussed not only statistical mechanics, but all of physics; and not only physics, but also social and political problems. Einstein heard us out with great interest. His personality was almost magical.

Perhaps 25 people attended the seminar. Besides Szilard, I recall two other Hungarians especially: one was a man named Kornfeld, who later returned to Hungary, took a Hungarian name, and entered politics. The other was Dennis Gabor, who later moved to England, became a superb physicist and engineer, and helped invent the electron microscope. In 1948, Gabor helped to invent holography, a kind of three-dimensional photography without lenses. And in 1971, Gabor was awarded a Nobel prize for his work with holography, one of the very few inventors ever given the Nobel prize. Jancsi von Neumann often attended Einstein's seminar, too, when he was in Berlin, so there was a strong Hungarian group within the seminar.

Szilard excelled at organizing groups of people along lines of common interest. And that talent for organization has always intrigued me. How many tactless people are gifted organizers? I have never met a single one besides Leo Szilard.

Szilard's circle often gathered informally on Saturday afternoons. Szilard's brother, who was not in Einstein's seminar, often joined us. We met in my room or in Szilard's. The family who rented Szilard his room allowed him the use of their living room.

Our meetings were friendly and casual but full of serious discussion. On Sundays, we read science books and papers in the library or joined other friends on museum trips. On fine spring days, we walked in the woods nearby.

Szilard supported my work stoutly. Once we both attended a meeting in Hamburg. Heisenberg gave a paper, and during the discussion that followed I offered an insight that Szilard dearly liked. He kept repeating, "See to it that your remark is printed!"

That none of this discussion was printed neither surprised nor much disturbed me. I was content to return to Berlin and resume my work. But it pleased me that Szilard wished to see my remark in print. I was a modest young man; I needed encouragement.

Szilard's reaction to Einstein's statistical mechanics seminar showed his deeply mixed feelings about physics. He was pleased with his role in urging Einstein to give the seminar; and what Szilard learned there apparently inspired his fine doctoral dissertation of 1922. He made strong proposals in the seminar and began a long and fruitful friendship with Einstein. Szilard often visited Einstein at home; later they even invented and patented a refrigerator together.

Yet I think the seminar also troubled Szilard; it seemed to rouse in him sour feelings toward higher mathematics. Complex abstractions rarely appealed much to Szilard, and he clearly did not relish the prospect of mastering a great many advanced mathematical laws.

I fear that the seminar may have convinced Szilard that he was not bright enough to change theoretical physics. And whenever Szilard felt he could not change something, his interest strayed. Mastery was not enough. So Szilard avoided most quantum mechanics, explaining to all who asked that it did not offer him enough.

Perhaps Szilard simply failed to see all the great work still to be done in quantum mechanics. Yet I fear he was chiefly hampered by mathematical unease. Szilard foresaw correctly that mathematics would become more central to quantum mechanics. But he was wrong to denigrate his own mathematical gifts.

Even with Szilard's distaste for higher mathematics, he might have written strong papers on the many philosophical issues which quantum mechanics implied. If he had been born 20 years later, he might well have chosen that path. But in the 1920s, philosophical papers were not yet respectable in physics. I tried hard to involve Szilard more deeply in quantum mechanics, even telling him that I needed a companion in my work. But I could not change his mind.

So Szilard drifted away from quantum mechanics. Besides his work for the seminar on statistical mechanics, he spent many hours doing x-ray experiments with my advisor Herman Mark. Mark had interests quite as broad as Szilard's, and together they also worked out some engineering ideas.

I am aware that I cannot be objective about Leo Szilard. But I must say that Szilard seemed very queer to me all of his life, and queer in essentially the same way. He had great talent, but talent hampered by an overly strong interest in himself. Selfishness is a very human property; it exists in all of us. But Szilard was selfish to an extraordinary degree. I wish I could explain why, but I cannot.

In the course of my long life, I have met a great many people. I have never claimed to gauge human character quickly, and many people have surprised me as I have come to know them. But given enough time and personal contact, I believe that I can nearly always understand a man: how he thinks; what he loves most; what he grasps and fails to grasp; and what motivates him to act as he does. The great exception to this rule, the only man who ever baffled me for a lifetime, was Leo Szilard. It may be more to my discredit than to Szilard's. But Leo Szilard is one puzzle I have never solved.

# Seven

# *Becoming a Physicist*

*A*fter taking my degree at the Technische Hochschule in 1925, I returned to Budapest. My father had found a job for me in the Ujpest branch of the Mauthner Brothers tannery. We worked there together most every day but Sunday. He did not supervise me directly, but we saw a good deal of each other. It was all just as he had planned.

I explained to my father the advanced tanning theories I had learned at the Technische Hochschule. He listened gladly, but to my dismay only partly understood. Specialized concepts are built on a foundation of basic knowledge; and much of the science underlying the tanning craft my father had never learned. Much of the knowledge that he had sent me to Berlin to master was knowledge that he did not have. It was hard to share it with him when I returned.

At first, I loved seeing Budapest again. It lacked the daily bustle of Berlin but for a time peace was just what I wanted.

And I assured myself that leather tanning was a worthy business. But over time, I became restless.

Relations with my father remained cordial and often far more than that. But I never felt quite at home in the tannery. I felt a certain repulsion for the way that some people were treated there. One day a man came in wanting to buy leather from my father. He gave some false information for which I mildly corrected him. At this he got furious and bawled me out. I was meek, but I had some pride. Being upbraided by a stranger did not appeal to me.

Most of all, I missed physics. Over time I realized that the desire for contact with physics was not something that I could control. It was a need. I tried to train my intellect on tanning principles, but questions about the virtues of various methods of tanning leather could not grip me as did questions about the structure of the nucleus.

I talked physics with some fine physical chemists working at the tannery, subscribed to the leading German physics journal *Zeitschrift für Physik,* and scrutinized the new articles by Werner Heisenberg, Max Born, and Pascual Jordan. I even joined a tannery club that discussed science and its relation to broad social issues. Still, it was not enough; I knew that I must be a physicist myself.

So I was delighted one day in 1926 to get a letter from a man named Weissenberg, a crystallographer at the Kaiser Wilhelm Institute. Dr. Weissenberg asked me to return there and assist him with x-ray spectroscopy, diffraction, and crystallography.

The offer was a genuine surprise. My degree was in chemical engineering, not in physics, and Weissenberg was a stranger to me. I soon learned that the offer was the work of the wonderful Dr. Polanyi. I badly wanted to accept. But I had more than my own wishes to consider.

I asked my father for advice. I knew very well how I should act while awaiting his answer: I should keep my own feelings hidden, giving him time to ponder the situation and allowing him to answer as he saw fit without undue concern for my inclinations. But I am afraid that I did not act that way. I simply could not contain the excitement I felt at the prospect of leaving Budapest and the tannery for Berlin and a life of higher physics.

This show of high spirits must have wounded my father, but he gave no sign of it. Quite the opposite; he kindly advised me to accept the offer. I think he understood then that I would never be the tanner that he had wanted for a son. His dream would not come true.

It had been many years since I had fully shared his dream, but how awkward it was to make that plain and cast off the dream for good. When I did so, my father was generous enough to forgive me. Hastily, I began preparing for my new job by studying crystallography.

I have continued to visit leather tanneries through the years, though I cannot say just why. Certainly I never thought of rejoining the craft. I suppose I felt a need to assure myself that I could still tan leather again if I had to. One never completely casts off the dreams of one's father.

When I reached Berlin, I vowed that from here on the study of physics would be my permanent occupation. It was a youthful vow, but sincere and deeply felt. My salary was miserable, about 450 marks per month, perhaps 140 dollars. Most of it went for food and rent. I was content.

But I knew that my parents would be disheartened if they knew of my choice to treat physics as a profession. It was with some pain that I realized I was learning things that my father would never understand.

I soon sensed how much I had missed while working at

the Mauthner Brothers tannery. Naturally, I had been following Heisenberg's pathbreaking work with half-quanta, but for all of my interest in his work, I somehow had no idea he was a man in his 20s. I thought Heisenberg must be some 55-year-old professor from Munich who had found a creative way to account for some experimental data.

I had subscribed to *Zeitschrift für Physik* in Budapest, but I saw that I should have also taken the competing monthly *Annalen der Physik* (*Annals of Physics*). At lunch with Polanyi one day at the Kaiser Wilhelm Institute, I mentioned my regard for the Born–Heisenberg–Jordan paper. Polanyi said, "Born–Heisenberg—that's a good beginner. But Schrödinger! Now, there's the one who really discovered things."

There was no mistaking the reverence in Polanyi's tone, and suddenly I recalled hearing many similar remarks about Schrödinger around Berlin. I confessed to Polanyi that I barely understood the meaning of Schrödinger's work. It was his artfulness that had engaged me so far.

Polanyi said, "What? . . . What?" Now, Michael Polanyi was the kindest man alive, a man who was never rude to a student. But he was genuinely surprised by my ignorance and did not try to hide it.

I made some meager reply, and the conversation moved on. But that "What? . . . What?" stuck in my ears. Immediately, I began to study Schrödinger's work with a new intensity. And after absorbing a good deal, I began to discuss it with Leo Szilard.

Right off, I could see that Schrödinger deserved all of the praise he had received. His work was a marvel. For all of its novelty, it still had a pleasing affinity for classical mechanics. Even today, I do not quite know how Schrödinger managed that.

My boss, Dr. Weissenberg, also did fine work. He wanted

to learn why atoms hold positions in the crystal lattice corresponding to symmetry axes and planes. He told me to read up on group theory, try to resolve this question, and then report to him. "Here is Weber's *Algebra,*" he would say. "Read it and then prove to me that stable positions in crystals are symmetry points."

I spent about three weeks reading Weber's book and I found a crude solution. Though Weissenberg was a fine crystallographer, he hardly followed my answer. He told me it was not general enough and sent me back to refine it.

This grew into a routine. Weissenberg gave me what seemed to be simple problems to solve. They *were* simple to solve in an elementary way. But then Weissenberg would look at my answers and ask for more elegant ones. Though I often doubted I could do better, the search for a suitable elegance led me increasingly deeper into group theory.

Group theory is physics with some axioms. The axioms tell you what the elements of the group can and cannot do: that every element of the group can be multiplied with every other element; that an "A" and "B," which are valid here, are also valid there. When you venture some way into it though, it becomes a great deal more complex.

I amused myself by trying to apply group theory to quantum mechanics. I did not get far, but I was very young and I thought that I had.

In group theory, Weissenberg gave me one highly exasperating problem. I worked diligently at it and got exactly nowhere. So I turned for help to my childhood friend from Budapest, Jancsi von Neumann.

Jancsi was then working at the University of Göttingen in Germany. He was devoted to exploring the implications to quantum mechanics of a concept called Hilbert space, a kind of multidimensional space that uses unit vectors to represent the functions of wave mechanics. Jancsi had been offered the chance to work in Göttingen with the man who had created Hilbert space in 1906, David Hilbert. So Jancsi had gone to Göttingen and helped Hilbert write a text expanding Hilbert space.

Even the prodigy von Neumann had not been left free to pursue directly the mathematical career for which he was so perfectly suited. Just as my father had given me chemical engineering in place of a physicist's career, Max von Neumann had given his son the field of chemical engineering.

Jancsi had spent two years at the University of Berlin reluctantly studying chemical engineering, then two years more in Zurich, where he had earned his diploma in 1925. Far more precious to him was the doctorate in mathematics that he had received in Budapest in 1926. The doctoral thesis and examination had scarcely troubled him. Quietly, Jancsi had dropped chemical engineering for his twin loves: mathematics and theoretical physics. He had also become something of a bon vivant. He enjoyed conversing with women, attending theaters and the cabaret.

I described to Jancsi the irritating problem in group theory which Dr. Weissenberg had just assigned me. Von Neumann understood it. He had not only the gift of cracking stubborn problems, but the far rarer ability to see their inner workings in perfect relation to any number of other, quite different concepts. Jancsi considered my group theory problem for about half an hour's time. Then he said, "Jenö, this involves representation theory."

Jancsi gave me a reprint of a decisive 1905 article by Frobenius and Schur. Von Neumann was both tactful and kind. He said, "You could perhaps work this out yourself, Jenö, but it's one of the things on which old Frobenius made his reputation. So it can't be easy."

Jancsi was exactly right; the derivations were not easy. So I turned to the article by Frobenius and Schur. Soon I was lost in the enchanting world of vectors and matrices, wave functions and operators. This reprint was my primary introduction to representation theory, and I was charmed by its beauty and clarity. I saved the article for many years out of a certain piety that these things create.

After this incident, I began to ask Jancsi more often to make calculations for me. "Jancsi," I might say, "Is angular momentum always an integer of $h$?" He would return a day later with a decisive answer: "Yes, if all particles are at rest." He was proud of his computational prowess and liked to handle complex computations for his colleagues. We were all quite in awe of Jancsi von Neumann.

After six months, I left Dr. Weissenberg to work with Richard Becker, a new professor of theoretical physics. Becker was a genial man of about 40. He knew a great deal about electricity and magnetism, both in classical and modern theory. My two semesters assisting Dr. Becker passed quite pleasantly. He required very little work of me, leaving me free to learn a good deal about electromagnetic fields.

One day in 1927, Becker told me, "I already know what you will do next year." Arnold Sommerfeld had written him, telling him to send me to the University of Göttingen as an assistant to David Hilbert. A man named Lothar Nordheim, a few years ahead of me, had been playing this role. But Nordheim was now off to Cambridge, England as a research fellow.

I had by now written a few physics papers. I doubt if Arnold Sommerfeld had read any of them, but he must have heard of them. Sommerfeld had been a theoretical physics professor at the University of Munich since 1906. He, too, was a pioneer in quantum mechanics—not as original as Dirac, Heisenberg, or Schrödinger, but a pioneer nonetheless. Sommerfeld studied the wave character of x rays and applied quantum theory to spectroscopy.

Happily for me, Sommerfeld also felt a duty to look after young scientists, both his students and others. Hans Bethe, who became a superb theoretical physicist, was then another protégé of Sommerfeld's.

The salary that Göttingen offered me to assist Hilbert was still miserable. But I cared nothing about that. What a splendid place Göttingen was for a young physicist.

The University of Göttingen was in western Germany. Göttingen was an ancient town where most people still traveled on foot. For more than a century, many of Europe's finest scholars had been coming there to work. Since 1880, Göttingen had been a center of science, not only in mathematics and physics, but in aeronautics and aerodynamics.

Karl Gauss, perhaps the greatest mathematician of the nineteenth century, had worked and taught at Göttingen until about 1850. Felix Klein, a superb mathematician and teacher, had taught there from 1886 to 1913.

Max Born had come to Göttingen in 1921, and had soon brought James Franck. Schrödinger had studied under Born at Göttingen and met Niels Bohr there. Heisenberg had come to Göttingen in 1921 with his teacher, Arnold Sommerfeld. Heisenberg had defended his own seminal paper on quantum mechanics there.

Past Göttingen students included Robert Millikan and Max von Laue. In a later generation, Karl Compton, Edward Condon, Norbert Wiener, Linus Pauling, and Robert Oppenheimer would all come. And though Jancsi von Neumann was no longer living in town, he still visited from time to time.

And in the three years before I came, Göttingen had pioneered the use of matrix mechanics in quantum theory: the assigning of an operator to every physical quantity, with the operators represented by matrices. That work excited me.

Most exciting of all was the chance to work closely with David Hilbert. Hilbert was probably the greatest mathematician since Karl Gauss. He had a towering mind and a manner that was somehow both gruff and quite inspiring.

Hilbert had been at Göttingen since 1895. He had developed both the theory of self-adjoint operators and a brilliant theory of invariants. Invariance involves the notion that the laws of physics must be independent of the location and movement of the observer. Hilbert had overwhelmed algebraic number theory with his gifts. He had even contributed Hilbert space to physics.

Does it seem odd for a mathematician like Hilbert to take a young physicist for an assistant? Well, Hilbert needed no help in mathematics. But his work embraced physics, too, and I hoped to help Hilbert somewhat with physics.

So I was quite excited to reach Göttingen in 1927.

I was quickly and deeply disappointed. I found Hilbert painfully withdrawn. He had contracted pernicious anemia in 1925 and was no longer an active thinker. The worst symptoms of pernicious anemia are not immediately obvious, and Hilbert's case had not yet been diagnosed. But we knew already that something was quite wrong. Hilbert was only living halfway. His enormous fatigue was plain. And the correct diagnosis was not encouraging when it came. Pernicious anemia was then not considered curable.

So Hilbert suddenly seemed quite old. He was only about 65, which seems rather young to me now. But life no longer much interested him. I knew very well that old age comes eventually to everyone who survives his stay on this earth. For some people, it is a time of ripe reflection, and I had often envied old men their position. But Hilbert had aged with awful speed, and the prematurity of his decline took the glow from it. His breadth of interest was nearly gone and with it the engaging manner that had earned him so many disciples.

Hilbert eventually got medical treatment for his anemia and managed to live until 1943. But he was hardly a scientist after 1925, and certainly not a Hilbert. I once explained some new theorem to him. As soon as he saw that its use was limited, he said, "Ah, then one doesn't really have to learn this one." It was painfully clear that he did not want to learn it.

I gauged Hilbert's plight at our very first meeting. After that, I saw him only about five times more the rest of the year. I did what little I could for him, but I could not make him younger or remedy his anemia. I had come to Göttingen to be Hilbert's assistant, but he wanted no assistance. We can all get old by ourselves.

❧     ❧     ❧

So in these first few weeks at Göttingen, I examined the course of my own life. I asked myself: What should I do? The answer to that question seemed quite clear. I should devote myself to physics. But that raised broader questions: Why is physics so important? What does present-day physics achieve? And what ought physics aim to achieve in the future?

Every scientist must ask himself these questions at some time, in some way. Most of us answer them quietly, to the satisfaction of our own souls. Often the questions are only asked and answered implicitly, in the scientist's work. But those first weeks at Göttingen were for me a time of deep and quite conscious reflection.

I had already learned the value of a life of physics. I was proud to possess that distinctly human trait that seeks knowledge that is not clearly useful. Useful or not, I decided that physics had a duty to provide a living picture of our world, to uncover hidden relations between natural events, and to offer us the full unity, beauty, and natural grandeur of the physical world.

For several weeks, I remained nearly alone in Göttingen. I had rented a single room. I was not in regular contact with Hilbert and had not received an introduction to anyone else. I was not a young man to put his nose in where he was not invited.

I found the mathematics library inside a kind of auditorium. I sat in that library reading physics texts and scientific journals and thinking as deeply as I could. A wonderful mathematician named Richard Courant came quickly over one day. Dr. Courant was about 40; he had taught mathematics at Göttingen for many years and was by then director of the local mathematics institute. After a brief, animated conversation, Courant said, "Mr. Wigner, we have a great deal more to discuss. But just now I am quite busy." And off he went. We saw each other again from time to time. In all the months I was in Göttingen, Dr. Courant and I never had the deep discussion about physics that I would have liked. Perhaps that was natural; our fields of work were quite different, and he was many years my senior. But I never forgot the simple act of kindness he showed me that early day.

I returned to my solitary thinking.

I knew that man needs something to strive for. I kept returning to that notion. What should the higher purpose of physics be? I decided that it should be to elevate the material side of the world, to make daily life easier for all the world's people.

This was more than the great classical works of physics had ever done. Isaac Newton's *Principia,* great as it was, had not immediately made human life easier. Newton's work had lived, had attracted support by its intellectual brilliance alone. But now, in 1927, I felt that physics was just one part of science, and science a part of technology. All three embodied human curiosity, and all three were working together at great speed to change the material world.

Our grandparents had lived full, prosperous lives without running water, electricity, or automotive power. Now, such lives were rapidly being left only to people we called "the poor."

Science and technology were making room for far more people on our earth, making their lives longer and easier, even giving them some leisure. By applying scientific knowledge to technical problems, we had nearly found the means to satisfy the material needs of all the earth's people.

And I felt this great movement of science and technology had done something subtler. It had raised the moral standards of human treatment. American Indians had often murdered entire enemy tribes. In Europe, whole peoples had been slaughtered, with little outside protest. This acquiescence to wholesale murder seemed to be ending—chiefly, I felt, due to the advance of the scientific movement. I felt that physics might soon begin to solve important human problems of a social and emotional nature.

So, I decided: If I could not work for Hilbert, I would work for physics.

As a rule, the older and younger scientists at Göttingen did not mingle. At first, I rarely saw any of the big shots. I was still an unpromising young man whose ideas jumped ahead of his writing skill and whose papers were rather obscure. But after a few months, I began spending a good deal of time with some of the younger physicists. I enjoyed the company of a theoretical physicist named Walter Heitler. Heitler was about one year my junior. He had recently become a Göttingen privatdozent, a young assistant professor, who is supposed to know how to teach. In later years, he did crucial work with theories of fast particles and cosmic rays.

Heitler was not a von Neumann or a Heisenberg, but he was a fine physicist. I must have spent more time with him than with anyone else at Göttingen. I teased him that he was a dabbler, but he knew how well I liked and respected him.

Heitler claimed good-naturedly that his work on chemical bonding would change the whole field of chemistry. Naturally, I was skeptical. I used to ask him, with a smile, "Well, now. What chemical compounds do you predict between nitrogen and hydrogen?" I knew that he did not have enough chemistry to answer me precisely. But Heitler was a very good man.

One day, I was lying on the grass near the Göttingen municipal swimming pool. Beside me sat the German astronomer Heckman. Suddenly, Heckman saw a lot of red ants crawling on one of my legs. He was surprised that I permitted this and asked did they not bite? When I answered that yes they did bite, Heckman asked why I did not kill them. "Well," I said, "I can't tell which ones are doing the biting."

I wrote a paper with Victor Weisskopf in Göttingen on the natural width of spectral lines. Weisskopf felt that a certain mathematical integral could be made to vanish. I showed him why the integral was infinite. The theory was imperfect then, and even now it is less elegant than it should be. But it served its purpose then, and has held up rather well in all the years since.

Victor Weisskopf did his thesis work largely with me. I also met the fine German physicist Walter Elsasser in Göttingen. Excellent physicists were always passing through.

Max Born and James Franck were two of the greatest

scientists at Göttingen—the same pair who had first published on the impossibility of an association reaction, after which Polanyi had deeply apologized for stifling this same idea in me.

Max Born was a friendly, thoughtful man and a founder of quantum theory. He could not yet have been 50 years old in 1928, but he seemed rather old to me then. Born had a strong feeling for the community of Göttingen. He did not stay there through World War II, but in 1953 he returned to live in semiretirement.

In 1927, Born was still intensely active. He and Pascual Jordan had co-written essential articles on quantum mechanics. With Niels Bohr and Werner Heisenberg, Born was one of the great philosophers of quantum mechanics.

I met Pascual Jordan and came to know him better than I did Born. Jordan had a superb imagination, but he seemed to feel that he could not compete with minds like Heisenberg and Wolfgang Pauli. I sensed that Born was the dominant partner.

James Franck was another giant at Göttingen, formal in manner but always cordial and a wonderful experimental physicist. It was Franck who confirmed that the maximum energy of an electron emitted by radiation obeys quantum theory.

So despite Hilbert's illness, the time I spent in Göttingen inspired me and left me even more determined to live out my life as a physicist. Göttingen seemed to me a paradise of physics.

# Eight

# "That Pesty Group Business"

$\mathcal{T}$he first year that I truly contributed to physics was 1929. Quantum mechanics had been evolving quite well without my help. So I wondered: What could I do? Where could I concentrate my effort and perhaps produce something original? I realized how much I loved group theory. I had worked with Michael Polanyi on the symmetries of crystal, and these symmetries had suggested group theory.

Group theory was a technique developed by mathematicians. It had been used by physicists years before I came along. But something kept occurring to me: that group theory had rarely been rigorously applied to quantum mechanics. Perhaps the dabbling I had done for Dr. Weissenberg might lead somewhere worthwhile.

There was nothing brilliant in this insight—just a bit of good instinct and much good fortune. I began working more seriously now, with the earnest willpower of an untested young man. Properly engaged in something potentially important, I found myself more patient than I would have thought possible. Carefully, I wrote a general equation for group theory and then set about applying it to quantum mechanics.

No one knew the full-scale application of that equation then; and even 65 years later, we hardly know it. Important consequences are never easily obtained. But that is the theoretical physicist's task: to carefully develop the consequences of things. And 65 years later, we see group theory as one of the best mathematical tools in atomic physics.

❖      ❖      ❖

In 1929, many physicists, especially the older ones, felt a certain enmity toward modern currents in physics. Many of them found group theory a nuisance because it treated physics as something stationary; they were used to regarding physics as motion. In the group theory of quantum mechanics, electronic orbits were now presented not as orbits but as spheres. Most older physicists disliked all this.

But I was young and the young are much freer of bias. The fact that group theory might be stationary and that its orbits seemed to be spheres did not irritate me at all. On the contrary, these facts rather appealed to me. Never having been trained otherwise, I readily digested the new formulations. I embraced group theory before most other physicists did, regarded it playfully, and wondered how it might advance physics.

Do you recall my description of Wolfgang Pauli and his genius for lively derision? Well, I believe it was Pauli who put a label on the common distaste for modern group theory. He called it "die gruppenpest"—in English, "the group pest" or perhaps better, "that pesty group business." "Die gruppenpest" became a popular label for a while.

Max von Laue disliked "die gruppenpest," though he was kind enough to support my work. Einstein considered all of group theory a mere detail. Even Richard Becker sometimes blurted out, "Die gruppenpest!" forgetting for a moment that he was a loyal supporter of his assistant, Wigner.

Erwin Schrödinger surely looked at group theory as "die gruppenpest." He wanted physics to develop in a particular way, and when it did not he was displeased. As late as 1934, Schrödinger said, "All this talk about the neutron is very boring." About my approach to group theory, Schrödinger told me, "This may be the first method to derive the root of spectroscopy. But surely no one will still be doing it this way in five years."

I relayed all these doubts to Jancsi von Neumann. We discussed the well-known gruppenpesters. Jancsi reassured me: "Oh, these are old fogeys. In five years, every student will learn group theory as a matter of course." And, as always, Jancsi was right. Soon all of them were learning "die gruppenpest"—even Wolfgang Pauli.

Max Born was another physicist dismayed by the growing respect accorded "die gruppenpest." But here, Born showed his integrity as a scientist. For despite his feelings, or perhaps to resolve them, he started a seminar on group theory in 1927, and took the trouble to learn the topic fairly well. My friend Heitler and I attended his seminar and spoke up quite often.

Another barrier to the acceptance of modern group theory was the lack of a first-rate textbook. The standard text on group theory came from a German mathematician named Hermann Weyl. Weyl was a superb thinker. He had formulated not only his own algebraic theory of numbers but also quite intricate ideas on the metaphysics of science. Weyl was about 15 years my senior, a former protégé of David Hilbert at Göttingen, and an old colleague of Einstein's in Zurich. Weyl had helped shape both quantum mechanics and the theory of relativity.

So Hermann Weyl thought very clearly, and his textbook, *Group Theory and Quantum Mechanics,* first published in 1928, had become the standard text in that field. Those who understood it saw in it a rigorous beauty. But Weyl did not write clearly, and so most physicists did *not* understand his book. Young students especially found the book awfully dense. For all his brilliance and good intentions, Hermann Weyl had discouraged a fair number of physicists from studying group theory.

Leo Szilard was the one who most encouraged me to write my own group theory text. He assured me that the field badly needed a book more accessible than Weyl's. Szilard also valued what he called "priority claims." When he invented a refrigerant, he placed a patent on it so that the world would know that the inventor was Leo Szilard. For Szilard, this was less a commercial claim than an intellectual one. Szilard felt that by writing a group theory text, I could establish a priority claim on the subject. He urged me to do so, and I agreed to try.

I worked very hard to write that group theory text, rephrasing my relevant published work and adding a good deal of unpublished work as well. The book took me two years; it seemed an infinity. But why not? I was an inarticulate young man with a partial understanding of an immense and beautiful subject. I wrote painstakingly, aiming for precision.

My little book was finally published in 1931. *Group Theory and Its Application to the Quantum Mechanics of Atomic Spectra* tried to make the methods of group theory clear to most physicists. I am afraid that my writing was poor; when Max Born sent me a list of the book's "errors," I found that most of them were misreadings of my argument. But through the years, many people have praised this little book. I am very grateful to Szilard for prodding me to publish it. And I am pleased and a bit surprised that I ever had the presumption to do so.

❖     ❖     ❖

It was at Göttingen in 1928 that I first met Paul Dirac. Dirac had already earned a reputation as the young master of quantum mechanics. He came one day by invitation to lecture about his recent work in quantum field theory.

Dirac's presentation was remarkable: rich in substance, but clear and detached, almost like a recitation of a technical text. Nothing at all was revealed about Paul Dirac. He was able to convey much of the beauty of quantum field theory, without giving any sign of enjoying his own lecture.

A casual discussion followed the talk. Dirac accepted questions readily enough but his answers, though quick and correct, were entirely restricted to the questions asked. Cleanly, but with an unmistakable firmness, Dirac declined to explore related subjects. He resolutely shielded from view all trace of his own opinions.

Soon afterward, I had the chance to take a few meals with Dirac at Göttingen, and I ventured to ask his opinion on various scientific topics. To these questions, Dirac responded in a more familiar tone. Still, he never spoke of his emotions or personal experience. I found this curious, and it was a few years before I understood why.

In late 1928, I returned to Berlin and the Technische

Hochschule as a privatdozent working with Richard Becker. I worked hard and wrote several successful papers for the Hungarian Academy of Sciences. My relations with Becker were quite unchanged. He had always liked to complain that promising young physicists were being distracted from their studies, and this I heard from him again.

I taught a class in quantum mechanics, but I made a poor lecturer. I knew the subject well and revered it. But I could not manage to convey that reverence to my students. One day my old teacher Richard Becker attended my class. I was chastened to see disappointment in his honest face.

I went to the great Wednesday physics colloquium and also to a Thursday colloquium on quantum mechanics. Schrödinger often was there, with Szilard, von Neumann, Heinz London, and some others. The colloquia often spilled over into a local coffeehouse.

We knew quantum mechanics then, but not as thoroughly as we needed to. It was not at our fingertips. Classical mechanics we knew at our fingertips: A rock thrown against a board strikes the board, makes a noise, vibrates, and the rock falls. We understood that process precisely. We did not yet know quantum mechanics that way. The papers I wrote with von Neumann were attempts to put quantum mechanics at our fingertips.

❧        ❧        ❧

It was around 1928 that I first came to know a man who would have profound effect on my life: Edward Teller. Leo Szilard, Jancsi von Neumann, Teller, and I were all born into Jewish families in Budapest within a single decade. Szilard was the oldest of this quartet, born in 1898; I came into the world in 1902, von Neumann in 1903. Teller, born in 1908, was the youngest.

I have already told how brightly von Neumann shone at the Lutheran gimnázium. I regret that I never knew Edward Teller or Leo Szilard as boys and have never known much about their inner lives as children. But I know enough to be sure that both Teller and Szilard were child prodigies. Of the Hungarian quartet, I was the slowest.

Szilard was also in Berlin around 1928. Von Neumann was at the University of Hamburg, developing precepts of game theory. Teller was in Leipzig, studying under Werner Heisenberg, who had just become a professor in Leipzig in 1927.

Heisenberg organized a big annual physics meeting. One year the meeting would be in Leipzig, and many physicists from Berlin would journey there. The next year, Heisenberg and his associates would come to Berlin. The meeting addressed pressing issues in physics.

At one of these meetings, I first got to know Edward Teller. We had met once three years before, but the contact was too brief to leave much impression. So in 1928 I was looking at Edward Teller freshly. What I saw was a young man of 20, six years my junior, of modest height, but well built and vigorous, witty and thoughtful. In attitude, he was not a youth at all, but a mature man.

Teller was studying theoretical physics problems quite different from mine, but he clearly enjoyed studying a great range of scientific problems. And he was from Budapest and spoke Hungarian, so I liked Teller from the start. Behind his pleasant, helpful manner I felt sincerity and warmth. And I soon learned that Teller liked taking walks and excursions. So we walked and took excursions together.

Teller's childhood had been affectionate. His father was an attorney. After gimnázium, Teller had begun his university studies at the Technische Hochschule in Karlsruhe, Germany, where he easily mastered mathematics, chemistry, and chemical physics.

The coming of quantum mechanics, which had seemed so miraculous to me in 1925, had equally delighted Teller. Eventually, it brought him to Munich to study with Werner Heisenberg and Arnold Sommerfeld. When Heisenberg had left for Leipzig, Teller had followed him.

The annual meetings organized by Heisenberg helped Teller and me to maintain a strong, irregular friendship. As I came to know him better, I saw deeper traits: kindness, loyalty, sound judgment, an intense dedication, and a rare capacity for understanding. I found that I liked Edward Teller dearly.

Teller was intensely curious about the world. But unlike many such people, his curiosity was never disagreeable. He did not pry. He was a bit like von Neumann in using a rich store of jokes to argue seriously about science and politics. Teller was modest, but purposeful. Despite his youth, he seemed to know what he wanted to support and achieve more clearly than I knew these things for myself. Time spent with Teller made you think.

Most of my talks with Teller were on the topics raised by Heisenberg's annual physics meeting. We disagreed on small, technical points, but never on fundamental issues. And yet I recall those talks as rich and subtle. There is a kind of magic at work in first meeting a man like Teller. You would have to be a very dull man not to feel it.

Often we pondered the quantum mechanics of Schrödinger and Heisenberg. Impressed on us by those times was the crazy idea that all of theoretical physics could be rebuilt on the implications of quantum mechanics. Teller and I wanted to be two of the builders.

Both Teller and I used to wonder how many Hungarians feared communism. It was a painful time, full of political violence. But most Hungarians were like the peasants working on my uncle's property near Belcsa-Puszta: They worked quietly, their political and social apprehensions unclear. My student friends worried far more about communism than fascism.

Teller and I were both quite interested in politics but our political discussions were few and brief. Rich political discourse is sustained by basic disagreement on such things as the proper role of governments and the degree of truth found in prevailing ideologies. On such things, Teller and I never disagreed enough to argue. We agreed that communist dictators might try to subdue the earth; halting them would be difficult, but crucial; people throughout Europe would likely have to be shown that the really beautiful things in life depend a great deal on political freedom.

I visited at least one museum with Edward Teller, and together we heard a chorus perform in a music hall. I was pleased to find that Teller loved poetry and music and played the piano with skill.

Once after a talk by Einstein on the Unified Field Theory, I visited a zoo with Teller and some others. I saw that Teller was feeling sad. Apparently, he had been unable to follow Einstein's lecture. I asked him what was wrong. Teller said sadly, "I am stupid." I considered that statement for a moment and then said, "Yes, that is a general human property." Teller quite enjoyed my comment. Because, you see, everyone before had always told him how smart he was. But I agreed with him that he was stupid, as we all are stupid compared to our ideal: That man recalls everything, absorbs novel truths instantly, and regards them with perfect judgment. That is not a human standard.

But even stupid people often make fine scientists. However wonderful a receptive mind may be, it is not central to science. Nor is mental speed. To persist in your inquiry and to engage it fruitfully—this is what makes a first-rate scientist. Curiosity, diligence, and ambition are traits far more essential than imagination. And stupid people are often remarkably well endowed with these traits.

❧        ❧        ❧

I also worked on symmetry in the late 1920s, teased by the idea that laws of nature have symmetries: If a law of nature works on a system constructed in a certain way, it will also work when the system is rotated or put in motion. I enjoyed being able to predict the development of symmetric systems.

My work in crystal symmetry suggests why scientists should follow their eyes and heart, worrying little about concrete applications for their work. When I began studying symmetry, most of those who knew me thought, "Well, that's a foolish thing." And I hardly disagreed. It *was* a bit foolish. But I did not mind playing the fool, because physical symmetry greatly pleased me. My favorite physical theorems might lack the full beauty of a great poem or the wit of a first-rate joke, but they had a special tricky charm.

If my work seemed trivial to some people, I did not care. Throughout my life, I have found it best to seek physics problems whose solutions seem initially simple. In complete form, their details revealed, such problems often become barely manageable. Solving physics problems that are exacting from the first often becomes a hopeless undertaking.

The exact theoretical consequences of my work in symmetry and group theory took years to develop. But eventually this work had quite fundamental applications. It spread the basic truth that laws of nature have simple invariance properties. It even influenced the basic design of nuclear reactors.

And the work had a wonderful immediate consequence: It convinced me that I truly belonged in the field of physics. The years between 1928 and 1931 were very rich ones for me. I learned an awful lot, worked very hard, and was very happy. It is a joyful thing to know that you are truly a physicist.

What else besides love can compare with it?

# Nine

# "If Hitler Says So, He Must Be Right"

*O*ne morning in Berlin near the end of October 1930, I received a startling cable from the United States: "Princeton University offers you a one-term lectureship. Please cable reply." It was nothing more than an offer to visit for six months. But I had never before received such an offer. And the salary they quoted was fantastically high. My salary in Berlin was then equal to about 80 American dollars per month, and I counted myself well off. I had just received the equivalent of 500 dollars for my book on group theory, and I thought that a handsome sum.

Now Princeton University was offering me about 4000 dollars, almost 700 dollars per month. I had never seen so much money in one heap. It was more than seven times my old salary. I could not conceive of such an amount. I knew that American professors earned more than Germans, that it would take an expensive boat trip to reach Princeton, and that prices were higher there. But not seven times higher. The salary quoted in the cable was so high that I felt it was an error in transmission.

I thought I knew why Princeton wanted me. The United States in those years was a bit like Russia: a large country without first-rate scientific training or research. Germany was then the greatest scientific nation on earth. I knew that many American universities wanted to improve in science. I thought at first that my physics work in Germany must have drawn the invitation.

Then I learned that Jancsi von Neumann had received a similar cable, with an even greater "error in transmission." Now I could believe the salary figure. And suddenly I knew quite well why Princeton had offered me 4000 dollars to cross the ocean. When two physicists of the same age from the same clan are invited at the same time to the same distant university, it is hardly a coincidence.

It was clearly Jancsi that Princeton really wanted. They had offered him about 1000 dollars a year more than me. And he fully deserved it. Though he was a year younger, Jancsi had already studied in Berlin and had taken both a chemical engineer's degree in Zurich and a mathematics doctorate in Budapest. While teaching in Hamburg, Jancsi had written mathematical articles so advanced that they ventured into physics.

Around 1929, Jancsi had married the former Marietta Kövesi. I had come to know Mrs. von Neumann well and regarded her with affection. Princeton did not want von Neumann and his wife to feel lonely in the United States. To be a stranger in a foreign country is a trying experience; far better to have an old friend around. And perhaps Jancsi would not have accepted a lone offer. So Princeton also invited Eugene Wigner. Promising as I might be as a physicist, it was clear to me that Princeton thought of me mainly as "the companion of Jancsi von Neumann."

You might wonder how Princeton learned about von Neumann and Wigner. Well, there was in those days a fine physicist named Paul Ehrenfest. Ehrenfest was 50 years old in 1930, a professor in Leiden, and a wonderful man. Einstein had taught in Leiden with Ehrenfest and greatly respected his work; together they had hatched the Einstein–Ehrenfest paradox which had so excited us at the colloquium in Berlin. Ehrenfest was not a founder of quantum theory, but he had applied the theory with great imagination and skill.

Paul Ehrenfest traveled a great deal. As a young man, he had found jobs hard to come by, partly because he was Jewish. As a result, Ehrenfest always liked to find work for young colleagues. In the 1920s, he began advising a few American universities on how to upgrade their scientific departments by hiring certain young Europeans. In 1930, Ehrenfest sent four German physicists to the University of Michigan. And he picked out two Hungarians for Princeton: von Neumann and Wigner.

Jancsi von Neumann and I had written three papers together in 1928 and two more in 1929. What a pleasure it was to work with von Neumann. I might be in Göttingen and he in Berlin. It did not matter. Each of us worked effectively alone. If I found a snag, I presented it to Jancsi. There was never a snag that he could not untangle.

He explained the most complex mathematical questions in a light, casual tone. If I told him I failed to understand Warring's Law, he might smile and ask:

"Do you know Hilbert's Third Proposition?"

"No," I would say.

"Then, do you know D'Alembert's Theorem?" he would continue, quite easily.

'     "Yes, I think so."

After three or four more questions, he would finally begin to explain Warring's Law, referring only to the theories that I knew and avoiding the others. By such circuitous paths, he quickly reached the core of the matter, which he explained easily. Von Neumann had the gift of making even the most complex concepts seem simple.

Jancsi von Neumann taught me more mathematics than any other of my teachers, even Rátz of the Lutheran gimnázium. And von Neumann taught not only theorems, but the essence of creative mathematical thought: methods of work, tools of argument. Much of this von Neumann had learned himself from Rátz, but most of it was his own. By 1930, Jancsi had well exceeded the mathematical reach of Mr. Rátz, as Rátz had foreseen he would 15 years before.

Jancsi was not only a genius as a thinker and a teacher, but also a splendid man and a fine companion. He was never conceited. He was quite unable to act pompously. Though he could be blunt when necessary, his habitual mood was one of a leisurely and graceful good humor.

He had grown more sociable since his days at the Lutheran gimnázium. No one knew more amusing anecdotes than von Neumann. His ability to coin a joke on any occasion delighted his fellows. Only intellectual dishonesty and misuse of scientific results could raise his ire. These did, whether he or another was the victim.

❋      ❋      ❋

I had ample time to weigh the offer made by Princeton University. But I hardly thought of refusing it. Six hundred dollars a month was a full professor's salary, enough money even to marry if I could find a rare and lovely girl.

Dutifully I told my boss, Richard Becker, and also Fritz

Haber, the director of the Kaiser Wilhelm Institute. Both men said, "Of course, you must accept."

Fritz Haber was quite an interesting man. He was a superb electrochemist who had won the 1918 Nobel prize in chemistry for the synthesis of ammonia from its elements, hydrogen and nitrogen. Haber had been director of the Kaiser Wilhelm Institute since its opening in 1911. After the First World War, he had tried seriously to collect gold traces from sea water as a means of paying German war reparations.

By 1930, Haber was well into his 60s, nearly bald, with spectacles and a large mustache. He was a gentle man but he liked to have a little fun. In my presence, he telephoned the Minister for Art and Science and told him of my invitation to Princeton. Haber added, "It's too bad the Americans have to tell us whom to promote."

I was later told that the Minister for Art and Science did not like that statement. But even he never doubted that I would accept Princeton's offer. Even my parents, who had always hoped I would stay near Budapest, agreed that I should go to America. They were flatly astonished by the offer of 16,800 marks for six months' work in a field they still considered dangerously unremunerative. I bought a boat ticket.

❧     ❧     ❧

Like most Hungarians, I knew almost nothing about the United States. I felt that it was a democratic, reasonable nation, with even some understanding of its original Indian inhabitants. But I had never expected to live there. I had a vague appreciation of American film stars. Some Hungarians followed certain American athletes, presidents, and popular songs. But I had not.

And even in 1930, I considered my move to the United

States only temporary. I was casual with the arrangements. I knew an American named Hubert Alyea who had recently received his doctorate at Princeton. I asked him, "Where is this place, Princeton, New Jersey? How do I get there? Does it have all the necessary stores? And where could I live in the town?" Dr. Alyea answered all my questions, and soon I was sailing for New York City.

The voyage took 11 days. I had studied English with a private school teacher in my last few months in Berlin. But brief training little helps a poor student. I passed the time on the boat struggling to read British editions of the novels of Thackeray and Sir Walter Scott. Finally, the boat reached New York Harbor.

New York City looked nearly crazy, but I liked it from the first. I saw that I would have to walk around a good deal before I felt acquainted. But I had always loved walking. I changed money and got about 30 of my first American dollars. Then I bought a railway ticket and rode to Princeton.

Jancsi von Neumann reached the United States about a day later. I met him, and we spoke Hungarian. We agreed that we should try to become somewhat American: that he would now call himself "Johnny" von Neumann, while I would be "Eugene" Wigner.

Eventually, other Hungarian friends came to the United States as well, among them Leo Szilard and Edward Teller. A Hungarian Club in New York even asked us to eat goulash with them from time to time.

I found quarters in part of the Princeton Graduate College, near the main university campus. The standard of living was far higher in the United States, though the rent was three times what I had paid in Berlin for a room no larger or more elaborate.

The town of Princeton had many fine walks. The soil was

poor by Hungarian standards, but then the people in Princeton did not seem to mind. They were not farmers here. It was rare to hear anyone talk seriously about the tilling of the soil.

Everyone seemed to own their own car in Princeton. Even the women knew how to drive and drove routinely. Despite this, Americans rarely bothered to visit the countries on their border, as we did in Europe. There seemed to be no interaction with Canada or Mexico. Americans settled more sparsely than Europeans and avoided their parents more. There was an air of dignified independence about family relations.

Princeton was a quiet town then, without the research institutes that have come in recent years. I had very few complaints with Princeton beyond its months of irritating humidity. After living in Berlin, I enjoyed a town where you know your neighbors and need not worry about stepping into the street and forgetting the name of the man approaching from the other side.

I was introduced to the other physics faculty. Howard Percy Robertson was friendly and asked me to call him "Bob." Oswald Veblen was courteous, and we saw each other occasionally.

I sensed quickly that the United States was a country where physicists revered solid work far more than they did in Europe. Here, ideas themselves counted for less and the long, hard calculus that developed them counted for more.

The Princeton physics department was hardly interested in quantum mechanics. Howard Percy Robertson followed it, but his interest was largely a formal one. The manipulation of the symbols of quantum mechanics was a game at which he was quite skilled. But he much preferred relativity theory.

The most important theoretical physicist at Princeton then was E. P. Adams, and his interest in physics was entirely macroscopic: classical mechanics, a bit of electrodynamics. There were a few others at Princeton who knew quantum mechanics, but most of them were mathematicians. And they knew it as an abstraction, not a precious link to personal experience.

I suppose my greatest complaint with the town of Princeton was the way that its people talked. The town had no coffeehouses in the European sense, where scholars and their students went for lively, extended conversation. And Americans spoke far too much English and not enough German or Hungarian. I was not foolish enough to expect to hear fluent Hungarian coming from the mouths of Americans; yet somehow I was dismayed to find that this beautiful language was almost completely unknown in New Jersey.

And the great culture that had grown from it was equally unknown. Even well-educated Americans had never heard of the Hungarian poets, Sándor Petöfi or Mihály Vörösmarty, or my teachers, Rátz and Polanyi. And then, all of my family was far away. So I felt as a stranger, a stranger in a friendly but quite foreign country.

Jancsi von Neumann regarded the United States quite differently. He had arrived with a greater reputation in physics and far richer job prospects. Having learned English from a tutor as a boy, he spoke it almost fluently. And he had a natural gift for small talk, which I conspicuously lacked. About Americans, von Neumann felt, "Here are a sane people, less formal and traditional than the Europeans and a bit more commercial. But a great deal more sensible too."

Jancsi felt at home in America from the first day. He was a cheerful man, an optimist who loved money and believed firmly in human progress. Such men were far more common in the United States than in the Jewish circles of central Europe.

Painstakingly, my English improved. My first talk at Princeton was at a colloquium in the chemistry department. I introduced myself as "Eugene." Determined to use my new English, I had memorized the first two sentences of my presentation. I hoped that after a smooth start I would relax and my new English vocabulary would come easily.

But after those first two sentences, I found myself quite stuck. For what seemed half a minute, I could not utter a sound. All the fine words of my English tutor, my new American acquaintances, of Thackeray and Sir Walter Scott had escaped me entirely. Finally, Hugh Taylor, a senior man in the department and a kindly fellow, said nicely: "Why don't you continue in German." I was surprised at my own vehemence: "No! I must learn English. I will continue in English!"

This exchange amused the colloquium guests, and the brief delay allowed me to regain most of my composure. I labored through my talk in what must have been a miserable English. But I made it. That was 60 years ago. Today, I can lecture quite easily in English. I can even swear in English.

I was not sociable in those first months. I did not want to intrude on the von Neumanns, who were newly married and needed privacy. I took a few short excursions with a young experimental physicist at Princeton named Charles Zahn. But little more than that. I would have liked to have dated a few women, but I never saw the chance.

That was all right; I was used to the solitary life, and it gave me time to examine more fully the workings of the United States. I found the American political system splendidly reasonable, about as good as could be made. The magic of democracy is that it never depends too much on any one man. Like any country, America has its rulers—powerful senators, great industrialists, and so on. As citizens, we enjoy complaining of their failings.

But these rulers do not control the very lives of people as the kings of Europe once did. Quite intelligent Hungarians used to live for 80 years without once even thinking of trying to install themselves as king. It was simply not done. Royalty lived off to themselves, as if in the clouds. In America, I saw many quite ordinary people set out to become president; some of them even succeeded. I found that quite heartening.

Clearly, the United States had its troubles. I was bothered by how poorly the Negroes were treated and by how firmly the races were separated. I quickly read some American history books and learned that the North Americans had also mistreated the Indians.

I had never realized in Hungary that the early white settlers in America had been conquerors like the British, the French, and the Russian czars. In some small way, I identified with the Red Man. I was intrigued that the Indian word "wigwam" was like my own name, Wigner; sometimes I even called myself "Wigwam."

I disapproved of both the cruelty shown the Indians and the seizure of their lands. But I felt, "Most nations have at some time in their history acted badly. Even the Hungarians had to take Hungary from its first inhabitants." I was sure that the American wars against the Indians were the better of the two kinds of war: an inevitable result of expansion.

❧     ❧     ❧

Apparently, Princeton was reasonably pleased with the work of von Neumann and Wigner. Just before the term was up, they made us an offer: For the next five years, why not spend half the year as visiting professors in Princeton and the other half year wherever you wish? The arrangement will benefit us both. Jancsi and I embraced the idea. We were not at all

Here I am, Eugene P. Wigner, around 1988. I have a weakness for reflection, and I want to leave some small record of the signal events of my life. (Photo courtesy of Eugene P. Wigner)

My wonderful parents, Antal and Erszébet Wigner, around 1930. (Photo courtesy of Eugene P. Wigner)

The faculty of the Lutheran gimnázium, in 1908. They had a presence. Rátz appears to be at the far right of the front row. (Photo courtesy of Eugene P. Wigner)

László Rátz, the first of my many great teachers. No one else could evoke a subject like Rátz. (Photo courtesy of Eugene P. Wigner)

Jancsi von Neumann, third from left, with his siblings and friends in 1915. Jancsi was truly a prodigy. (Photo courtesy of the AIP Niels Bohr Library)

The Lutheran gimnázium, around 1919. I am the second boy from the right, in the front row. It may have been the finest high school in the world. (Photo courtesy of Eugene P. Wigner)

Albert Einstein. All of the rest of us were in his shadow. (Photo courtesy of Burndy Library)

Paul Dirac as a young man. Two things hurt his reputation in Berlin: he was too English and he lacked emotion. (Photo courtesy of the AIP Niels Bohr Library)

Erwin Schrödinger. In 1926, he recast quantum mechanics in his own startling artistic spirit. (Photo courtesy of the AIP Niels Bohr Library)

My first wife, Amelia Frank. When she passed away in 1937, the love which had blessed us began to seem like a curse. (Photo courtesy of Eugene P. Wigner)

Leo Szilard. It was hard being his friend, but it was always deeply interesting. (Photo courtesy of the Argonne National Laboratory)

Enrico Fermi. He had a perfect willingness to accept facts and men as they were. (Photo courtesy of the Argonne National Laboratory)

The bottle of Chianti with which we toasted the first controlled nuclear chain reaction in man's history. (Photo courtesy of the Argonne National Laboratory)

A Wigner family portrait, taken around 1950. Mary Wheeler Wigner, Eugene P. Wigner, and our two children. (Photo courtesy of Eugene P. Wigner)

Accepting the Nobel prize in physics in 1963. I had never expected to get my name in the newspapers without doing something wicked. (Photo courtesy of Eugene P. Wigner)

Paul Dirac near the end of his life. His departure from this world made a widow of my sister Manci and brought me very great sorrow. (Photo courtesy of Eugene P. Wigner)

Johnny von Neumann. All of his very great achievements arose from a single coherent view of life. (Photo courtesy of Alan W. Richards)

Returning to Budapest around 1975, for the first time in some years. (Photo courtesy of Eugene P. Wigner)

Discussing physics with my good friend, Edward Teller. Old Hungarian friends never completely lose touch. (Photo courtesy of Eugene P. Wigner)

Edward Teller, in a recent photograph, one of the very few which reflect his wonderful sense of humor. Too few people appreciate Teller. (Photo courtesy of Edward Teller)

Standing with my third wife in front of our house in Princeton. Eileen Hamilton Wigner is a loving, trusting companion and a constant friend. (Photo courtesy of Eugene P. Wigner)

ready to abandon our teachers, friends, and family in Europe. But we could see that American physics was improving and we were glad to have a place in the United States as well. We appreciated the generosity of the offer. And the Technische Hochschule seemed to notice Princeton's quickened interest. They gave me a teaching appointment for the other half of each year.

1932 was a miserable year for the world's economy. It was a wonderful year for physics. An Englishman, James Chadwick, discovered a wonderful subatomic particle called the neutron, and soon thereafter Heisenberg wrote his classic paper stating that the nucleus consists of protons and neutrons. As so often happened with Heisenberg's papers, the details were later superseded, but the basic ideas were sound and inspiring.

In 1933, the situation for Jewish scientists in Germany quickly worsened. In January, President Hindenburg invited Adolf Hitler to become chancellor. Before 1933, I had never thought seriously about Hitler or the meaning of his rising political career. I must have seen his photograph, but none of my friends attended his rallies or heard him speak on the radio. Very few Hungarians did.

Hitler's vile speeches were printed in German and Hungarian newspapers, mostly Nazi papers. They attracted little attention. It was a time of great hardship in Germany, and hundreds of crackpots promoted their own programs for delivering Germany from its enemies.

We vaguely knew about this man Hitler who made brash speeches about the glorious future of a purified Germany. We knew vaguely that he had gained many followers and that these followers had begun to call themselves "the wave of the future." But we underestimated Hitler.

Adolf Hitler was an easy man to underestimate. He

looked perfectly all right: a little man in dress uniform, with a plain German face and a small mustache. There were many German men in 1933 who looked very much like Hitler. But Hitler was not all right. He made himself first a terrible nuisance and finally a vicious dictator.

Adolf Hitler must have had some goodness in him somewhere. But I have never known anyone more purely evil. There may have been other people as wicked as Hitler, but we did not usually let them acquire any real degree of power. Hitler was allowed to seize enormous power, and so his evil had enormous manifestations.

Hitler expected to conquer the earth and made no secret of it. Just as the various parts of Germany had been joined to make one strong country, Hitler felt that all the countries on earth could become one entity, ruled for 1000 years by Nazi Germany. Hitler was very clever. His speeches boasted: "The Germans are the right people, the best people. We will rule the world." To rule the world: What a heavy idea that is! Years before Hitler came to power, he was promising that Germany would rule the world. His promises were not logical or consistent. But neither are human emotions. So Hitler's appeal was dishonest and dangerous but also quite effective.

Hitler's political rivals could not say, "No, this is false. We Germans are really no better than any other peoples." That would have been the truth, but in national politics the truth is often less popular than a clever lie.

And Hitler also played on the strength of German fear of communism. He promised that his Nazi party would physically resist the communists in the streets, and this reassured many Germans. They gave Hitler surprising backing in the Parliament and he quickly expanded his power.

Most Germans seemed strangely unconcerned with Hitler. They had not sought him, but when he came they said,

"Well, the man is impressive. Let us see what he does." Few of them expected him to lead Germany into a disastrous war. They watched with interest as he blamed their hardships on Jews and on other nations. But most of them thought he would stop short of a war. After all, Hitler had managed to take power in Germany without a war.

And if Hitler did start a war, most Germans felt that they would win it. Hitler had often bragged of it and had never been clearly contradicted. Hitler had always been very successful. He had the rare gift of complete self-assurance. When a man believes in himself completely and transmits that belief, when he always knows what to do today and easily vanquishes his rivals, people assume he will prevail in the future. So when the Nazis called themselves "the wave of the future," many people decided that indeed they were.

Hitler always seemed to know what to do. He had made himself the sole dictator of a country as great as Germany, and had done it very quickly and easily. So the Germans said, "Well, if Hitler says so, he must be right. We accept his actions." Soon they *had* to accept Hitler's actions, whether they liked them or not. Hitler was not a man who asked permission. That is the way it is with dictators.

Hitler knew something very important: that people need some kind of creed to strive for. They need to love something and many of them need also to hate something. I know very well that most civilized people reject the idea that many human beings have a need to hate. But I mean that quite seriously. The 1930s was a decade of terrible, fundamental mistakes. The worldwide inflation and unemployment made clear that our leaders did not grasp the basic nature of money and the workings of modern national economies.

As the Nazis rose to power, it seemed clear to me that most people also discounted the human hunger for power and hatred. That human beings can hunger for hatred is a very ugly idea. I dwell on this only because it is terribly important. Our recognition of this truth was bought very dearly.

To love, Hitler gave the Germans the idea of a mighty German nation. To hate, he gave them the Jews. He encouraged people to speak openly of an idea that had been mostly hidden beneath the civilities of European culture: the idea that the Jewish people should be entirely annihilated, with all of their culture and every trace of their time on this earth.

The dream of eliminating the Jews is a very old dream. Various forms of it have lived for many centuries. To people of other religions, Jews have always seemed selfish and disloyal. Most Christians through history have misunderstood Jews and at least vaguely resented them.

Adolf Hitler's Jew-hating was far more evil and literal than the ancient kind. Hitler felt that no one born Jewish could ever be a "real" German, even after a religious conversion. Hitler planned a sweeping revolution against "Jewish subversion." He wanted to degrade every Jew in Germany.

In February 1933, Hitler suspended freedom of speech and freedom of the press. By April, he had begun firing all the Jewish or semi-Jewish people in the universities, promoting the "real" Germans, extending his influence among them. And by October he had removed Germany from the League of Nations.

Hitler split Germany deeply between the "real Germans" on one side and "the foreigners, the impostors, and the Jews" on the other. As a foreigner of Jewish ancestry, my annual visits to Berlin were clearly numbered. That was clear, just as clear as it is in September that in December will come a chill.

Hitler's campaign against the Jews cost him most of the greatest people I had studied with. My old teacher Michael Polanyi accepted the rise of Hitler with a sad grace. In 1933, he resigned his life membership in the Kaiser Wilhelm Society and his position at the Kaiser Wilhelm Institute. He moved to Manchester, England, and began a new life. He wrote, traveled, and studied patent law, philosophy, and economics. He did all this very successfully, but I doubt he was ever again as happy as he had been in Berlin.

Leo Szilard also moved to England in 1933. Dennis Gabor left his research position with a firm in Berlin in 1933 and moved permanently to England. Richard Courant went to Cambridge, England in 1933, and then on to New York. Walter Heitler went to Bristol, then to Dublin and Zurich.

After 22 years at the head of the Kaiser Wilhelm Institute, Fritz Haber left for Switzerland, his spirit broken. Within two years, he was dead of heart disease. Erwin Schrödinger went to Graz, Austria, and in 1938, barely escaped to Oxford, and from there to Belgium and Dublin.

Victor Weisskopf, who had been working as Schrödinger's assistant, went to Kharkow, Russia, then on to Copenhagen, to Cambridge, to Zurich, and finally to the United States. Lothar Nordheim, who preceded me in Göttingen, moved to Paris, then to Holland and the United States.

Seven important professors were removed from Göttingen in the spring of 1933. Max Born was one of them; he left for Cambridge and then Edinburgh. Born's partner, James Franck, left Göttingen in protest, for Copenhagen and then Baltimore.

Walter Elsasser moved quickly to Zurich where Wolfgang Pauli was already well established. In the years since I had first seen Pauli at the colloquia in Berlin in 1921, he had studied in Göttingen, Hamburg, and Copenhagen. He had been a full professor in Zurich since 1928.

Pauli at first felt rather safe from the Nazis there. Zurich was in neutral Switzerland, and only Pauli's father was Jewish. Pauli had been raised by his mother as a Catholic. But by 1940, Pauli, too, had left Europe for the Institute for Advanced Study in Princeton. He did not return to Zurich until the Second World War ended.

Albert Einstein renounced his German citizenship and also moved to the Institute for Advanced Study. Hermann Weyl did the same. Herman Mark waited a few more years, then left Vienna in 1938, for Canada and then Brooklyn, where he had a brilliant career making polymers and finding the structures of cellulose, rubber, and silk.

Paul Ehrenfest took his own life in Leiden in that bitter year of 1933. Ehrenfest had serious troubles besides Adolf Hitler, but I think it is fair to say that his great heart was overwhelmed by the evil of the Nazis. I joined the general scattering.

If Hitler did not personally fire us all from our jobs, he made life unsafe for us in Nazi countries. And he forced us to follow politics. Many other important physicists fled as well, Enrico Fermi, Edward Teller, and Hans Bethe among them. By 1941, most of us had resettled in the United States. So perhaps the United States should erect a great stone monument to Adolf Hitler for his dedication to advancing the American natural sciences. Not even Joseph Stalin scattered scientists like Hitler.

From 1930 to 1933, Jancsi von Neumann and I oscillated together between Princeton and Europe. Once I spent six pleasant months in Manchester, England at the invitation of Michael Polanyi.

I tried to spend my summers with my parents at their country house 20 kilometers north of Budapest. There I got to see my sisters again, with their husbands and children. Time with family becomes far more precious when you are separated half the year by a great ocean. Time spent in Berlin was precious in a very different way: It allowed Jancsi and me to keep in touch with many of the world's greatest physicists: Einstein, Schrödinger, von Laue, and the rest.

But the situation for Jews in Germany rapidly became intolerable, and it dawned on me that Jancsi and I would soon be unable to return there. My "temporary" move to the United States had become something far greater than that. I would now need a permanent home outside Europe. I hoped against hope that fascism would subside and Hitler be replaced or subdued. But I did not expect it to happen.

I knew that many Germans respected the Jews, and I was appalled that they let Hitler do as he pleased. But, you see, my years in Germany had been spent among scientists, a great many of them Jewish. Even an "Aryan" like Werner Heisenberg was surrounded by Jews. Jews were his best friends and collaborators. Jews sat with him at conferences and in coffeehouses. Heisenberg never thought to protest the great prestige of Jewish physicists. He was shocked to see a German government abuse them. But not every German was a Heisenberg.

By 1933, I saw Europe as a sinking ship. I felt that I was certainly a European and belonged on that ship. I still looked forward each year to the months when I could return home to my family and friends. I still subscribed to a German newspaper.

I felt guilty abandoning the sinking ship, however much safer life might be abroad. And so, as I often did in times of doubt, I turned to Jancsi von Neumann for advice. I found him as pragmatic and self-assured as ever. He was also intent on staying in the United States. In a sense, Jancsi had fallen in love with Princeton on the very first day. We both knew that leaving Europe forever would mean turning our backs on a great deal of tradition. But that kind of break never troubled him as it did me; he was far less emotional than I was, far less dependent on tradition.

Jancsi, with his great realism, saw that there are times in this world when even great traditions lose all of their meaning. And Hitler had succeeded in rendering many of Europe's greatest traditions perfectly meaningless. Jancsi said simply that it is folly to treat with nostalgia what is irretrievable. I began to agree.

Von Neumann asked me a simple question: Why should we stay in a part of the world where we are no longer welcome? I thought about that for weeks and came up with no good answer. Yes, the ship of Europe was sinking. But I realized that the Nazis did not consider me a part of the ship at all; knowing that made it possible to abandon the ship.

In 1933, Princeton University got a new president, Harold Dodds, and President Dodds seemed well disposed toward the sciences. I felt rather secure at Princeton, and was quite conscious of how lucky Jancsi and I were to already hold teaching jobs in the United States.

I wanted to help other professors forced out of Germany. An old aquaintance from the University of Berlin colloquia, Rudolf Ladenburg, had come to Princeton in 1931. Ladenburg and I tried to find money and teaching posts for German refugee professors. Many of them were desperate for jobs. Luckily, the United States needed scientists. Ladenburg and I were not as helpful as we had hoped to be, but we did some good.

You know, it was not only Jews that Hitler killed. He also murdered the mentally retarded. The Jews Hitler accused of being overly ambitious. To accuse the mentally retarded of overstrong ambition was too big a lie even for Hitler. So he accused them of not being ambitious enough.

Hitler claimed Germany needed brains to conquer the

earth and rule it properly; the disheveled brain of a retarded man would only hinder Germany's destiny. Hitler had the retarded suppressed, even murdered. Then Hitler saw the gypsies, and he said, "See how the gypsies make music against me! They too must be killed!" That is the kind of man Adolf Hitler was.

In the years since 1945, people have often said, "Dr. Wigner, how perceptive of you to foresee the threat posed by Adolf Hitler!" Such praise disturbs me. Looking back at myself in the 1930s, I see a distinctly *unperceptive* young man, absorbed in learning physics as best he could, wanting no part of national politics or warfare. It did not take any special perception to see the Nazi will to subjugate. It took a special perception *not* to see it.

Just after the war, a series of photographs documented the gas chambers in which Hitler had asphyxiated millions of Jews. Around the world, people were deeply shocked at the photographic images of wholesale murder. I was not shocked. Gas chambers were just the technical means of a hatred that Hitler had always shown very clearly to anyone who wished to see it.

Harshly and abruptly, Hitler ended the flow of my pleasantly reasonable life in 1933, as he did the lives of millions of others. By the end of that year, I felt the Jews of Europe being swirled into danger, as though by a terrible whirlpool.

For the first time I saw how right my father had been 15 years before when he and I had discussed the nature of man. My father had quietly insisted then that human desires have no proper rationale and that no moral law is inviolable. Now I saw with some bitterness that he was right. Adolf Hitler and the Nazis demonstrated very clearly that the only absolute laws are laws of science.

The next 12 years were not reasonable ones.

# Ten

# *It is Far Better to Have a Good Marriage Than a Quarrel*

*I*n 1934, my sister Manci joined me on my half-year sojourn to Princeton. I felt that she needed a new environment. Her husband was a tall, good-looking fellow with a fine Hungarian name, but Manci had recently separated from him, and for good reason. I was concerned for Manci. I felt that I could offer her support and diversion, and I believe I did. But I did not foresee how much support Manci would offer me. Her presence was calming.

My single room was clearly too small for Manci to share. I asked the von Neumanns if they might put her up, and they were pleased to do so. The von Neumanns had a substantial home on Westcott Road in Princeton. They gave elegant parties, attended by German and Hungarian émigrés. At the von Neumann's house, Manci had her own situation and seemed to enjoy herself greatly.

❖　　　❖　　　❖

Between 1931 and 1933, Paul Dirac often visited Princeton University for a few weeks at a time. During these visits, we had become friends, and because neither of us had many friends, we found ourselves taking meals together almost constantly. It was then that Dirac first hinted to me of his unhappy childhood.

He had been born in Bristol, England just a few months before I was born in Budapest. Like me, he was one of three children and had received special attention from the better teachers in his high school. But unlike me, Dirac had resented the strictness of his father, who was a French instructor and apparently overfond of discipline.

Like me and Jancsi von Neumann, Dirac had wished to study purely mathematics and physics. Like us too, he had been given a more practical field by his father. For him, it was electrical engineering. By 1931, Dirac no longer studied electrical engineering, but he clearly savored what it had taught him. He used to say that the laws of nature are approximate, but the uses and limitations of approximation are quite subtle.

Two things about Paul Dirac struck most of the scientists from continental Europe: first, that he was English, and second, that he lacked all emotion. Neither trait recommended him. But I liked Dirac almost from the first. There was much in his upbringing that I identified with and something in his manner that greatly appealed to me.

Gradually, our friendship deepened. Dirac was not a man to say, "I feel now that we are truly friends!" He was quite reluctant to confide. The very act of speaking seemed to pain him, to summon memories of a sad childhood. He used to say, "There are always more people who want to speak than to listen." So Dirac made himself a listener.

Dirac was right: There *are* more people who want to speak than to listen. But most listeners will eventually speak up when conversation idles. This Paul Dirac did not do. He liked to listen even when no one was speaking. He wanted company at meals, but silent company.

Dirac's behavior went beyond modesty. He kept apart from the world and his fellows, confining himself to his own brilliant mind and strong convictions. And yet his temperament was one that pleased me. Before long, I knew very well that I liked Paul Dirac dearly. And I think he knew it, too.

One evening at a restaurant, Manci looked over my shoulder and said, "There is a man in the back who is very surprised to see us." She asked me to turn and look back at him. Turning around, I was quite pleased to see Paul Adrien Maurice Dirac. When I told Manci that Dirac and I knew each other well and had often eaten together, she told me to invite him to join us. This I did, and Dirac accepted the invitation. I formally introduced Dirac to Manci.

Dirac never told me his thoughts at that dinner. Perhaps he forgot them himself. But it was a very important dinner. Dirac had no direct knowledge of love then. But as he came to know Manci, he fell in love with her. Science has never been able to explain why Dirac should fall in love with Manci and not another woman. It took me about four months to realize what had occurred in that room.

When I saw that Manci loved Dirac equally much, I was terribly pleased. I felt that I had been right to bring Manci to Princeton. And it showed me two welcome sights: Dirac as a completely human person and Manci as a quite happy woman.

Manci returned to Hungary soon after meeting Dirac, but before long Paul visited her there and likely met my parents. Paul Dirac and Manci were finally married on January 2, 1937. Dirac became a loving husband and adopted Manci's two children. And together they had two more children.

Manci had never much understood physics and what she

had understood of it she regarded with a firm dislike. In all the years of their marriage, Manci could never bring herself to embrace physics. Paul did not mind; he knew that however she felt about science, Manci was quite fond of him. And Dirac managed somehow to remain very fond of both physics and of Manci.

❊    ❊    ❊

Princeton made me a full-time teacher in 1935–1936. They called me a visiting professor. The appointment was an important one which others had sought. I lectured and spoke at colloquia. I taught very little. I was always pleased to teach about the application of group theory to quantum mechanics. Though I was not a natural teacher and my English was still miserable, my prospects for staying at Princeton seemed excellent.

Princeton's level of physics was then quite rudimentary. Its physics students were considered among the best in America, but then American high schools had far more pupils than the European ones I had known. In many ways, this was admirable. But it meant that the American prodigies who entered Princeton had received far less extra teaching from gifted teachers than they would have in Hungary.

So we at Princeton often had to teach what ought to have been absorbed in high school or in a coffeehouse. Coming straight from the cultivated physics environment of Berlin, I often felt in Princeton that I was talking baby talk. Still, I had brilliant luck with my own graduate students. And in the 1930s it was far easier for professors to have intensely personal relations with their graduate students.

My very first graduate student was Frederick Seitz. Seitz later joined the Manhattan Project, developed a theory of solids, and eventually became president of the National Academy of Sciences. Fred Seitz understood the needs of people, just as well as he understood the cohesive energy of metallic sodium. Seitz took his doctorate in 1934 and left me, but we have kept up through the years, talking not only of physics but of the nature of social and political dilemmas.

My second graduate student, John Bardeen, had an even more eminent career. He took his doctorate in 1936, did pioneering work in solid-state and low-temperature physics, helped invent the transistor, and was awarded a Nobel prize for it. Bardeen became the first man ever to receive two Nobel prizes in a single field when he took the physics Nobel again in 1972 for his work in superconductivity.

And my third student was also superb: Conyers Herring, who went on to a brilliant career as an applied physicist. Herring was a researcher at Bell Laboratories for over 30 years and has since been an outstanding university professor. So, my graduate students at Princeton were better than I deserved. It was deeply satisfying to work closely and on one subject with such young men. Fred Seitz especially I felt was in many ways more mature than I was.

Another man whose work meant a good deal to me in these years was Hans Bethe. We had met just once, briefly, in Manchester around 1933. But Bethe's work I knew quite well. He wrote a marvelous series of articles in *Reviews of Modern Physics* in 1936 and 1937. They were really unbelievably good. I had a slightly younger colleague who liked to mark up many of these papers, noting Bethe's "crazy mistakes." But I was ravished by Bethe's work. It synthesized beautifully nearly all of what was known in nuclear physics at that time.

❧    ❧    ❧

Like me, Albert Einstein came to Princeton as a small experiment, but found he liked the town and the university and decided to stay. Technically, his job was at the Institute for Advanced Study. But we also made him a member of the Princeton University physics department. The department had only about 12 members in 1933 and few of them were Europeans. So Einstein and I had much in common; we came to know each other far better than we had in Berlin and far more nearly as equals.

Sadly, Einstein was far less close to his faculty colleagues than he had been in Berlin. And he had almost no contact with students, either graduate or undergraduate. There were several reasons for this. First, Einstein never felt at home with English. For "E squared," he always said, "E *quadrat.*" Now, "quadrat" may be intelligible, but it is not English. Einstein knew that, and it made him a bit uneasy with English speakers.

Second, Einstein was now no longer chiefly a physicist. He was preoccupied with Hitler's plans to subjugate Jews and conquer the world. Perhaps he was less anxious about this than I was, but he was concerned in his own profound way.

Third, Einstein's interests in physics diverged from those of almost everyone else at Princeton. His great dream was to modify the General Theory of Relativity, to make it form a common basis for all physics, and perhaps even for all science. But most physicists then had far humbler goals: To apply quantum mechanics to the theory of atoms and molecules, to the properties of metals, and to the principles of chemistry.

Also, the Princeton colloquia were nothing like those in Berlin. Instead of reviewing three or four papers weekly, the Princeton colloquia gave just one, always written by the speaker. This meant that many major papers were omitted, isolating Einstein further from mainstream physical thought.

Einstein's interests remained wonderfully broad. He wrote lively German poetry. He admired the great philosophers, thought like one himself, and gave his letters a philosophical tone. His literary reading was not terribly wide, but he brought to it great understanding. He loved the music of Bach and Mozart, and delighted in playing his violin.

But Einstein's friendships were confined to a small circle of collaborators and friends from past days. I was lucky to be one of these. I knew Einstein for more than 20 years in Princeton.

I could not think quite at Einstein's level, but then no one could. But I had a number of qualities that must have appealed to him: I was a physicist and a European; I was a minor link to his glorious past in Berlin; I was a pleasant-enough man; I spoke a very solid German; and I greatly enjoyed walking. Perhaps most important, I was one of the few people near Einstein who loved much the same physics he did. I treasured Einstein's faith in my favorite symmetry and invariance principles and often told him so.

We often walked together, discussing the wonders and bothers of physics. Einstein did not tell jokes in any formal sense, but his lively humor colored his thoughts, even on points of science that were quite technical.

One of the true privileges of Einstein's friendship was the chance to hear him describe powerful ideas that he had never published. You might expect Einstein to publish every idea that he knew to be important. But no; he often chose to withhold brash ideas that seemed to contradict larger theories of which he was fond. For instance, he never published his idea of the guiding field, for it seemed to him to defy the conservation laws of energy and momentum.

Einstein was a world-famous genius and people I knew used to remark, "You spend a good deal of time with Einstein. He has a perfect brain, doesn't he?" Well, I have never known what is meant by a "perfect brain." I do know that Einstein's mind was very human and had many defects. Einstein was far slower than Jancsi von Neumann to derive mathematical identities. His memory could be faulty, at least after 1933. And he was hardly interested in the details of physics. For a man like Edward Teller, developing the details of a physics problem was passionately important. For Einstein, it was not. In all spheres of life, Einstein's greatest pleasure was in finding, and later expressing, basic principles.

But Einstein's understanding was deeper than even Jancsi von Neumann's. His mind was both more penetrating and more original than von Neumann's. And that is a very remarkable statement. Einstein took an extraordinary pleasure in invention. Two of his greatest inventions are the Special and General Theories of Relativity; and for all of Jancsi's brilliance, he never produced anything so original. No modern physicist has.

I never knew Einstein's emotions closely. I doubt that anyone did. He had traveled a great deal as a younger man, but as he aged, he rarely ventured far from Princeton. He used to sail his beloved boat on a nearby lake and walk every day from his home on Mercer Road to the Institute for Advanced Study. But he kept mostly to himself. Never once did I see him lose his temper.

Einstein often seemed to me lonely, and yet I cannot say with any confidence that he was. For there are many known remedies for loneliness: the making of new friends, chattering with old friends on the telephone, and so on. And these things Einstein never did. So I think that he was not a lonely man but by nature a solitary one, who did not want his weaknesses to show, and did not want assistance when they did.

Einstein was divorced from his first wife around 1915; his second wife died in 1936. He never spoke to me of either of his wives, of their lives together or his feelings for them. He did not seem attracted to women in the way that most men are.

Einstein was not a good family man in the conventional sense. He was friendly to his family and devoted to them in his own way, without being truly close to them. Somehow in a man as gentle and profound as Einstein, the habit of emotional distance did not seem a fault. One simply observed that Einstein was no more like other men as a husband than he was as a thinker.

I doubt that Einstein regretted his flaws as a father or husband. It was enough for him to think about physics and about great human problems. While most men were thinking, "Now, where is my wife? And what shall we have for dinner tonight?" Einstein was wondering, "Oh, why are there Nazis in this beautiful world?"

No man is completely free of prejudice and self-interest —not even Einstein. He generally acted with a detachment akin to serenity. But the very word "detached" implies that a man is aloof, and anyone who says that Einstein was aloof did not know him very well.

Einstein certainly loved children and perhaps my favorite memory of him involves my own children. Around 1950, my wife took some pages of my physics work to Einstein's home one day. He asked her about our small children and she had to admit that they had the chicken pox. Local health regulations had forced her to leave them in the car. Einstein said, "Oh, I have had chicken pox already. Seeing them for a moment surely will not hurt me." And he proceeded to walk down to the car and have a long talk with my children, which they can still recall. I doubt that Einstein even knew what chicken pox was. But he knew what children were.

❧     ❧     ❧

In 1936 came a shock: Princeton dismissed me. My appointment had apparently aroused the jealousy of others who felt they deserved my job. When my office expired in 1936, some of them convinced Princeton not to reappoint me.

Princeton was quite a different university then: prouder,

more self-important, old-fashioned, and isolated. Besides Einstein, it had a few other fine modern thinkers. Oswald Veblen, for example, was a quite progressive mathematician.

But the Princeton Physics Department did not think normally about the facts of life. They felt superior to the Physics Departments at Columbia University, the University of California, Berkeley or the University of Chicago. They barely kept in contact with any other universities. The University of Göttingen had also been a bit this way, but with a great deal more justification.

You see, the scientists at Göttingen were ahead of their time. The Princeton Physics Department was behind the times. In 1934, few of Princeton's physicists fully accepted quantum mechanics. By then, most European physicists saw that quantum mechanics gave a wholly consistent explanation of all atomic and molecular phenomena below the speed of light.

Erwin Schrödinger and Paul Dirac had just shared the 1933 Nobel prize for their work in quantum mechanics. Werner Heisenberg had received a Nobel prize too. So it was clear that even the Nobel committee now admired quantum mechanics and found theoretical work worthy of the highest recognition. Yet Princeton had not yet embraced quantum mechanics.

The Princeton Physics Department never explained why they did not reappoint me. They just did not. They did not say, "You must go away," but they did not rehire me. Some group of professors must have felt I did not fully deserve my job. So they said, "Let's get rid of him. We didn't know him when we asked him here. Von Neumann is far better than Wigner, and von Neumann is now well settled."

Apparently, I did not impress them. And I began to feel that all along von Neumann and I had been treated as nothing more than two extravagant European imports. Karl Compton, who was then chairman of Princeton's Physics Department, was a famous physicist and quite likely a fine department chairman. But when he left to become president of MIT, Jancsi and I had been at Princeton six months and Karl Compton still could not tell us apart. That fact stuck in my mind.

I could not help feeling angry that my colleagues would get rid of me this way. But what could I do about it? It was not clear to me that I was a good physicist. It is quite hard for a young physicist to judge his own worth. You love physics and you work on it, but how skillfully is rarely clear. Certainly, I had made no great innovations. So I felt that perhaps Princeton had been right to fire me.

What I resented was their method. If I was misbehaving, they should have told me directly. If I had been unable to improve and had to be fired, it would have been courteous to inform me as soon as the decision was made and to help me find a new job in this foreign country. But no one did any of these things.

I began seeking a new job. I have said that in those days American physics was growing and needed well-trained professors. So it should not have been too hard to find work. But these were the days of the Great Depression. Universities were poorly funded and decent jobs hard to find. I was not far from despair. I told von Neumann what had happened. He was no longer at Princeton University himself; when the Institute for Advanced Study had begun in Princeton in 1933, Jancsi had been invited in.

The Institute was a great experiment in American higher learning and research. For a 30-year-old Hungarian mathematician to be asked into its first ranks was quite an honor and made Jancsi's integration into American life complete. So, his situation was far more secure than mine.

Jancsi listened carefully to my story, but he did not react

as most people will to a tale of deceit. A mind as inexorably logical as von Neumann's had to see and accept many things that the rest of us do not; this colored his moral judgments. Jancsi loved axioms and he used to say, "It is just as foolish to complain that people are selfish and treacherous as it is to complain that the magnetic field does not increase unless the electric field has a curl. Both are laws of nature."

As a child, von Neumann had been brilliant and unrestrained. But as he matured, he grew to see the consequences of his brilliance and the effect of his words on others. At times, he tried to restrain his mind. As I told him that I had been fired, I felt that he was restraining his mind in this way. Von Neumann could do little more than promise to recommend me for jobs where he could.

Next I wrote to Gregory Breit, a theoretical physicist at the University of Wisconsin. Breit had once worked at the Institute for Advanced Study, and we had written a well-known article together on the spectra of chemical reactions.

Gregory Breit was a curious man and an odd person to appeal to for help in my situation. He was an intense, thin-faced Russian immigrant who wore spectacles and liked to speak German. Breit was unruly in his enthusiasms, almost addicted to his physics work. And though I liked him, many of his associates did not. Breit did not follow standard social norms, and when he was aroused, he had a violent temper. He was as sparing with praise as any man I have ever known.

Why then did I turn to Gregory Breit? Partly because I had few choices. I knew just a handful of important scientists in the United States, and Breit was certainly one. But there was something more than that. Somehow I felt that Breit could make an effective advocate. He was quite loyal to the few men with whom he had worked successfully, and I was one of these. To me, even his wildness had a distinctive, friendly tone. Breit was tenacious in pursuit of what he thought was right. And for all of his abruptness, his wildness, Breit could be terribly careful when he chose to be.

In his own mind, Gregory Breit was always devoted not only to science but to the greater goal of human helpfulness. Breit's trouble was that he was highly intelligent and just as intense, far too passionate for polite society. But in his own way, he helped a great many of his colleagues.

So I was greatly pleased but not completely surprised when Gregory Breit persuaded the University of Wisconsin in Madison to offer me a job as acting professor in 1936. I never discussed with him what arguments he made on my behalf or why. The subject would not have been an easy one for either of us. I accepted the position on the spot.

It has always delighted me that it was Gregory Breit, a fighter, an untamed and often disliked man, who helped me find this job at a time when I was badly shaken. The world is often curious that way.

❋     ❋     ❋

The state of Wisconsin charmed me from the first. This, I soon decided, was the real America, where the common people were open and friendly, grew potatoes, and knew the simple life. In Wisconsin, I met farmers. In Wisconsin, I saw great wheat fields and the hard, honest work that went into their care. All of these things were a part of my happiness. But the main part of my happiness was the generousity of the other physicists in the department. They brought me right into the group. They made me feel at home from the first day. Everyone treated me as a friend. A group of young professors brought me to the university gymnasium and together we ran around in circles.

The University of Wisconsin was rapidly becoming a center of nuclear physics research. But the professors there still worked freely with the outer world. They were not in the clouds; they had friends outside academic circles. These professors helped the new and less gifted teachers. Their conversation addressed not only physics, but politics and social problems. They never seemed to fight for tenure.

A physicist named Ray Herb was the one who really kept the department together. He was about five years younger than me and had just recently arrived in Madison. But Ray was a great enthusiast about physics and life, enormously unselfish and tireless. He seemed to work day and night, and the whole department was infused with his spirit.

Life in Wisconsin made me realize that I had been unhappy at Princeton for months. And inside, I thanked Princeton for having fired me. Let me be clear on this subject: Being fired is unpleasant. I do not recommend it. But sometimes it is the best thing that can happen. If I had never been fired, I would not have seen the University of Wisconsin. I would not have met many fine people or seen a whole new region of the country. It was at Wisconsin that my deepest love of this country was born. In Wisconsin, I truly became an American.

I kept working with Gregory Breit. We developed quite an intriguing theory of neutron absorption in 1936. We wrote two joint articles in 1937 and another in 1938. We told *Physical Review* a bit about the capture of slow neutrons and dispersion formulae. Together we created the Breit–Wigner Resonance Formula, which has held up rather well through the years.

And at Wisconsin I met Edward Creutz, who later worked closely with me both in Princeton and during the Manhattan Project, and went on to a superb scientific career.

❊　　❊　　❊

I enjoyed the company of my colleagues immensely, but there are some relations that a man cannot supply. I wanted female companionship. I was neither young nor entirely innocent. But I was heartily romantic. I came to Wisconsin missing deeply the idea of a faithful, romantic love.

Before too long, I met a lovely lady in Madison. Amelia Frank was a young Jewish woman, though religiously not very observant. She was a physics student at the university, who had worked with the esteemed physicist John Van Vleck, and was considered a highly promising scholar.

It was not long before I realized that I was in love with Miss Frank. I wondered why, surprised not by her beauty, but at the fact of love itself. Romantic love is not a human need like food or shelter. The survival of the human race requires no tenderness of feeling. So I wondered: Why do people fall deeply in love? I found no clear answer. But I resolved that one day science should try to fully explain love.

In Hungary, men were supposed to marry at 30. My father had done so. I was now almost 34; it was time that I marry too. Amelia Frank and I were united in marriage on December 23, 1936.

My wife was a lively, healthy woman. But within a few months she had pain and then fever. It was soon clear that she was seriously ill. We learned she had heart trouble though we never knew the exact terminology. She may have inherited the condition. We did not know; her family were nearly strangers to me.

I called a doctor that I knew. His diagnosis was indefinite. I think he feared to tell us that she would soon pass away. He must have seen that he could not cure her. But he made no predictions, and for that I am grateful. Better not to learn what the doctor cannot know for sure; the premature diagnosis of my childhood "tuberculosis" had weighed heavily on my family.

My wife rested in bed, and with a bit of simple treatment she seemed to revive. The doctors advised her not to join me on a proposed trip to Europe in the summer of 1937. They did not want her to travel, but they pronounced her much better.

Then her condition worsened and she had to enter a hospital in Madison. She stayed there for about two months. Near the end, she briefly visited her parents' home. I was still pretending that she was essentially healthy. I thought it best to try to conceal her impending death from her.

But she knew death was near; she told me so. And on August 16, 1937, my beautiful wife Amelia Frank Wigner left this world forever.

My grief was long and intense. If she had been sick when we married, neither of us had known it. Nine months of marriage is not nearly long enough to know each other as a husband and wife should. But we had become quite close emotionally. The love between us, which had seemed an undeserved blessing, now felt like a curse. I was terribly lonely. I recalled part of a poem I had learned from Michael Polanyi. It can be translated this way:

> To my dark life I obtained a sweet light
> Now the sweet light's extinguished
> I am surrounded by night

We are all guests in this world.

My life was now complicated and seemed to me unjust. I drifted along at Wisconsin for some months, hoping time would fade my grief. But it did not, and eventually I felt I should find a way to leave a place that I would always associate with Amelia Frank.

By 1938, Princeton needed a new physicist. They wanted John Van Vleck, the man who had been a mentor to my wife. Van Vleck was a physicist a few years my senior. He was brilliant. Many years later, he was awarded a Nobel prize. Princeton might have gotten him in 1934, when he had left Wisconsin for a job at Harvard University. But by 1938, Van Vleck was content at Harvard.

Princeton decided that they wanted to hire a rising young physicist willing to stay on there for many years. They asked John Van Vleck to recommend someone of this description. Guess who he suggested? Eugene Wigner.

So Princeton invited me back to nearly the same job they had dropped me from two years before. I would never have accepted their offer but for the death of my wife. But now life in Wisconsin had become too painful. I resigned from Wisconsin on June 13, 1938, and in the fall became a professor of mathematical physics at Princeton.

The Princeton physics department hardly blushed at having dismissed me, and I felt no bitterness in returning. I doubt that anyone apologized for my dismissal; but then the invitation itself was a kind of apology.

I settled again into life in Princeton. To my great pleasure, the university seemed truly intent on creating a powerful modern theoretical physics department. They no longer wanted baby talk. And yet the advances made in the department had a real cost. There were more graduate students and our relations with them were far less intimate.

Life in Princeton had changed too because Jancsi von Neumann's marriage had broken up in 1937. Jancsi never told me why; he may not have known the reason himself. The von Neumanns seemed happy; they had certainly hosted many gay parties in Princeton. They were brilliant entertainers.

But marriage is not always an entertainment. Jancsi could be a bit distant from the world. He liked fine cars, but he drove them poorly and wrecked a few of them—not because he did not know how to drive safely, but because he hardly bothered. In a world full of intriguing mathematical puzzles, Jancsi could not see how the driving of an automobile deserved his full attention.

Perhaps he also slighted his family. But that is just a distant guess; what ends a marriage is never easy to know. Jancsi never asked my advice on any aspect of married life. I doubt it ever occurred to him to do so. And if he had, what could I have advised him anyway? That it is far better to have a good marriage than a quarrel?

# *Becoming Pleasantly Disagreeable*

*B* ack in Princeton, I watched closely for news of Adolf Hitler. I wondered how the Hungarian people regarded Hitler now. I knew that his Jewish suppression would not cause much fuss; most Hungarians were not fond of Jews either. But as Hitler made clear how badly he wished to suppress Hungary itself, I wondered: When will my Hungarians finally awaken? From my vantage point in Princeton, it was very hard to know. But most Hungarians seemed to genuinely like Hitler. I found that remarkable.

More remarkable still was that even the Jews themselves hardly seemed to hate Hitler. They knew he was unreasonable and they said so. But is that hatred? The objection was an intellectual one, and hatred is not a matter of intellect but of deepest emotion. Jews in Germany hardly listened to Hitler. When he had proposed their suppression in 1930, many of them had said, "Oh, this is just wild talk so that he comes to power more quickly."

When Hitler took power and began practicing his cruelty, the Jews *did* get angry. Many of them wanted to depose Hitler. But did they really hate him, as he hated them? Again, I think not. Hatred is not easy to define, but to me hatred means that if you saw the chance, you would murder whom you hate. Very few Jews would have seized the chance to murder Adolf Hitler. I suppose I might have done it. But my parents would not have, nor would anyone else I knew. They would only have deposed him.

But, you see, Hitler really did hate the Jews. He despised the Jews and he murdered them as soon as he saw the chance. To speak of the Nazi crimes rouses bitter memories. But I feel that the subject should be raised regularly to help prevent it from recurring.

Adolf Hitler killed about 6 million innocent Jews. About half a million of them were Hungarian. Murder on that scale is very hard for me to understand and very hard to forget or forgive, even after 50 years.

❖　　　❖　　　❖

In 1934, my summer months in Europe were strained by the quickened rise of the Nazis. My relations with Richard Becker were just as they had been five years before. Becker was still the same man, kindly finding work for aspiring scientists and complaining that trivial matters lured the most promising from their studies.

Other friends were more concerned with politics. On June 30, 1934 came what was called the "blood purge." Ernst Roehm, chief of staff of the Nazi paramilitary group the Brown Shirts, was killed by the Nazis. If Adolf Hitler had not planned the deed himself, he was thought to have approved it. Roehm was an important Nazi organizer with strong ties to the German Army. He was a difficult man, of independent mind and strong temper. But as far as the public knew, he had been a loyal ally of Adolf Hitler for many years.

Several other important Nazis were killed at the same time: General Kurt von Schleicher, the Reich Chancellor; and Gregor Strasser, head of organization in the Nazi party, a man who had recruited thousands of followers to the banner of socialism and believed deeply in its promises.

Berlin buzzed with the news of the June 30 murders, but on every point opinions differed. Roehm's personal goals had never been well known. The politically initiated spoke knowingly of plots and counterplots. Simpler men skimmed nervously over it all, intent on their own troubles.

I felt sure that these executions would leave lasting footprints on Germany. And my friend Rudolf Ladenburg strongly urged me to leave for Princeton. I realized with a chill that things were no longer safe in Budapest either. It would be hard for me or Michael Polanyi to relax now anywhere in Hungary. Frank discussion of politics had become a luxury in much of Europe.

In September 1934, I returned to the United States on board the ship *S. S. President Harding.* It was a troubling voyage. I was shaken by the thought that I must take my trips to America far more seriously now.

Eleven Rockefeller scholars were on the boat, including two mathematicians headed for Princeton. I talked shop for a time with one of them, but eventually the discussion turned to politics. Both of them had extensive political opinions, but though they condemned the Nazis in a general way, they would not speak with complete honesty.

The ship's crew was largely German and partly Nazi, though even Nazis were more reticent on an American ship bound for New York than they would have been in Berlin. The arguments of the Nazis seemed not only cruelly distorted to me, but childish.

I was reading Thornton Wilder's novel *The Bridge of San Luis Rey* aboard that ship, and even the novel disturbed me. I was troubled by Wilder's treatment of death and contented myself with *South Wind,* a hefty, more pleasant novel by the popular British writer Norman Douglas.

⚜ ⚜ ⚜

In January 1937, I became a naturalized United States citizen. Already, Hitler's allies in Hungary were taking the land and possessions of Hungarian Jews. Nazis liked to steal from Jews. They certainly would have taken the precious country home that my father had found for us, but I think the Communists had seized it already. So, too, my uncle's country estate near Györ and the wonderful estate near Belcsa-Puszta where I had first learned to love human speech.

Even turning Lutheran could not ensure the safety of a Jew. Nazi beliefs on religion were certainly not sophisticated. But they clung to one unshakable tenet: that a Jew was a Jew and could never make himself a Christian, just as a Frenchman could never make himself a German nor a man make himself a woman.

This belief that "a Jew is a Jew" had been quite general in my youth, but since the First World War many Jews had made smooth conversions to Christianity. Then Hitler came along, insisting that any Jewish ancestors made you unalterably Jewish, and being Jewish made you fit for any kind of abuse.

The lies Nazis told about Jews were incredible. It was like the braying of donkeys. But these lies were hardly challenged.

I knew that I was likely safe from the Nazis in America. But I saw no sense in sitting on a high moral level, out of sight and out of reach. I felt strongly that Jews in America had a duty to speak out, to affirm their Jewishness, to protect themselves. What, after all, is the essence of democracy but every group struggling to assert its own interests? I could not shrug off this duty.

My parents would clearly have risked their lives by staying in Hungary. I helped them get their immigration papers and brought them over to the United States around 1939. By then Germany was already threatening to invade some of its smaller neighbors.

My monthly salary of 550 dollars brought my parents to the United States. It helped them to survive Hitler and to see that perhaps the scientific career of their son had not turned out badly after all.

Emigration is quite stimulating to a young person. But my parents were no longer young in 1939. My father was 69, my mother about 60. Neither of them had ever seen the United States before and neither spoke a word of English. Suddenly, they had lost their language, their culture, their friends.

My father had been working at the Mauthner Brothers tannery for about 50 years. He held a respected position there and was used to being considered quite important. He wanted to find a similar position in an American tannery. But at age 69 and without any English, he could not. It is very hard to learn English at age 69, especially if your wife is Hungarian.

So my parents did not work. They lived in Princeton at first, then moved to rural New York State. My mother had never held a formal job in Hungary, but she had always kept busy maintaining the house and arranging their social life. There was less such work for her to do in America.

I never knew just what my parents hoped to find in the United States. Little, perhaps, but some happiness and the company of their children. Well, they found their children. But we were not enough to fully support their happiness.

We tried to cheer them. I reminded them of my affection and admiration. I made sure they got American newspapers and dictionaries. I introduced them to some Hungarians and Germans in Princeton, but most often the age difference was too great to support a friendship. In later years, after I had remarried and had two children, I showed my parents their little grandchildren.

But my father especially lost interest in life. The United States considered him a quite unimportant man. And that was a great and painful shock to him, which I could barely soften.

Most children know how much they depend on their parents. Parents see less clearly how much they need their children. It was hard for me to cheer my parents. My father especially could not bear to rely on his children.

So, even though my parents knew they were safer out of Hungary, they never appreciated the United States as I did and as I had hoped they would. They never really embraced this country. I had hoped they would come to adore a land where they could live without suppression. But they could not. They spent their time wishing that Hitler had never existed, and that they had been allowed to remain in the familiar comfort of Budapest. Happiness is not easy for people in their 70s. I understand that much better now.

The first great event for nuclear physics had come in 1932. It had been entirely experimental: James Chadwick's discovery of the neutron. In 1938 came another such discovery: Otto Hahn and Fritz Strassmann, working at the same Kaiser Wilhelm Institute for Physics in Berlin where I had worked 15 years before, split the uranium nucleus and discovered nuclear fission. Their achievement followed work done by Enrico Fermi's group in Rome. But Hahn and Strassmann made a splendid improvement; together they made it clear that nuclear fission would soon be able to be controlled and used for man's own purpose.

Before the discovery of fission, people did not expect to ever see atomic bombs. Very few physicists even expected it. A famous British physicist had said that the crucial thing to remember about nuclear energy was that it could not be used. And most of us agreed with him. The day that Hahn and Strassmann discovered fission, this began to change.

Now, let me sketch Leo Szilard's deeds after leaving Germany in 1933, because the next few years saw his greatest work. Szilard was likely the first person to see the real danger of an atomic bomb. It was years before the value of this vision was recognized, and it is still not widely understood today.

People like to label one man or another the "Father of the Atomic Bomb." A foolish exercise, for the atomic bomb had no single father. But we can safely make Leo Szilard one of its wise uncles.

In the spring of 1934, I visited Szilard in London. As Rudolf Ladenburg and I had done in America, Szilard had started a group to help non-Aryan scientists fleeing the Hitler regime. But Szilard needed money himself.

You might think a brilliant young scientist with ties to Einstein would have had many job offers. But in England, Szilard was a foreigner and a Jew. He did have some fine industrial patents, but he was not yet a well-known scientist. And he was a willful person with uneven manners and very little tact. So Szilard struggled to find a job which he thought worthy of him.

Independent of Enrico Fermi, Szilard had already seen the efficiency of collisions between neutrons and nuclei. He saw the chance for a neutron chain reaction creating a nuclear bomb, though the original basis for his conviction later proved false. Because he was Leo Szilard, he held to the general idea tenaciously.

We discussed this idea at length, and Szilard showed great insight. He tested the idea with experiments in 1934. On March 12, 1934, more than four years before the splitting of the uranium atom, Szilard applied for a patent on the laws of nuclear chain reaction. A 1935 patent application to the British Admiralty gave detailed calculations and plans, which proved invaluable within a few years.

I did not expect any of Szilard's patent mechanisms to produce nuclear energy, but I watched them with great interest. Controlled nuclear energy seemed only a question of time. I told this to a man from the General Electric Company in late 1935 who consulted me on other business. And I said as much to friends.

In the spring of 1935, I gave a short lecture to a popular discussion society in Madison, Wisconsin and predicted we would have nuclear energy production in five years. But I had very little basis for that number. And I did not expect to play any vital role in producing it.

Never once had I thought of working for the government. Why should I? I was a physicist, not a politician. I did not wish to be ordered around or to direct others. But I was shaken by Hitler's plans to subjugate Europe and perhaps the whole world.

Hahn and Strassmann's discovery of nuclear fission in 1938 suggested that the first atomic bomb would likely come from Nazi Germany. Hitler could conquer the world far more easily with such a bomb. Soon I began thinking of working with the United States Army against Hitler. And I tried to get other scientists to offer their services to this cause.

The discovery of fission by Hahn and Strassmann had an even more dramatic effect on Szilard. He felt that now, finally, his fears had been justified. Here was the missing key, the proof that a huge nuclear bomb was possible.

❈        ❈        ❈

In September 1938 came the Munich Agreement. Hitler had his eye on the part of western Czechoslovakia called the Sudetenland. About 3 million people of German origin lived there. British, French, German, and Italian officials met in Munich. They solemnly agreed that peace could be maintained if the Nazis were allowed to annex the Sudetenland. The Czechs were not invited to the meeting.

Neville Chamberlain, the British prime minister, came home from Munich promising "peace in our time." I knew very little about Chamberlain or the country that had produced him. But the "peace in our time" statement seemed to me not only foolish but quite dangerous. Leo Szilard knew a good deal about England, and he loved that country. But after the Munich agreement her future seemed to him dubious. Soon he moved permanently to the United States.

Sadly, as late as 1938, most Americans hardly noticed Hitler's plans of conquest. If you described Hitler to an average American, he would say, "What a hateful man! If what you say about him is true, he will someday be overthrown by his own people. And if he tries anything like that in this part of the world, he will be abolished." But if you asked just when and how this could occur, the American would say, "Relax. We have plenty of time. Germany is very far from here, and we whipped her in the First World War. Her army will never match ours." Most Americans cannot believe that their lives need ever be touched by developments in Europe.

Even my friend Edward Creutz felt this way, and Creutz was a man of great intelligence and sense. When he left the University of Wisconsin in 1939 with his physics doctorate, I invited him to Princeton, eventually convincing him to work on developing a chain reaction. During the war, Creutz helped make the nuclear reactor at Hanford, Washington.

But at first even Ed Creutz did not like the idea of working on the bomb. I think it was Creutz who told me: "Eugene, you are pleasantly disagreeable to press me into this project." That was the general reaction to my appeals: that I was pleasantly disagreeable to ask people to sacrifice their time and pleasure just so that America might build a wild new explosive with which to threaten the brash ruler of Germany.

I disliked being "pleasantly disagreeable." I took a certain public stand in urging America to war only because I felt it was imperative. And I could never blame the Americans for resisting the idea. Good people dislike waging war. My own family had recoiled from the horrors of the First World War. The urge to ignore Hitler in the 1930s was natural.

None of the Frenchmen I knew in 1938 thought Hitler would really overrun France, even as he lavishly promised and prepared to do just that. The French assured themselves that France was a prominent nation, with a strong military and sound social convictions. The French argued that German soldiers, even if pressed into battle, would not fight them wholeheartedly.

The French were fooling themselves. They had no thought of France ruling the earth, so they had no idea how badly Hitler wanted to rule the earth and how much he was willing to sacrifice in the effort. France could not bring herself to imagine all this.

Indeed, very few people in the 1930s built their ideas about Adolf Hitler on a foundation of reason. The great majority, even among the well educated, began with certain convictions about Hitler and then, more or less skillfully, found reasons to justify them. I already knew that belief is not rational at heart, but I was surprised to see how badly people wanted to cloak their beliefs in reason.

It was refugees from other parts of Europe who saw most clearly that a war was coming. We reported the menace to our adopted countries. Both in England and in the United States, we did all we could to herald the danger and to alert those in power.

In the process, we learned something important about human nature: that people do not like to learn from foreigners. They prefer to rely on their own friends and countrymen, those with whom they have grown up and feel an instinctive kinship. As foreigners, we knew Americans far better than they knew us. We were surrounded by Americans and had spent years absorbing their culture. Americans rarely had any meaningful contact with foreigners and hardly understood us.

Even as late as 1940, America was a strange mixture of people: a very few who had seen the danger of Hitler since 1933; a majority group just beginning to see it; and a sizable minority who still felt that the United States was in no danger at all. These stark divisions in political opinion made it a time of striking conversation, which I would have enjoyed far more had I not been so anxious about Adolf Hitler.

Compared with either Edward Teller's or Leo Szilard's, my political views were moderate. But as late as 1940, my politics made a great many Americans label me "a European extremist." These people never said just what they wanted the "European extremists" to do. But they seemed to feel we should forget all of the strangeness of Europe and think instead of how blissful American democracy was, how beauteous her wheat fields, and so on.

The antiforeigner feeling was never violent, but it was disturbing. By 1939, my mentor Michael Polanyi had settled down in Manchester to help England begin its sad and difficult task of resisting Hitler. The problem could not be avoided, and Polanyi worked closely with others to fashion England's defense. Polanyi, too, met antiforeigner feeling in his work.

❖      ❖      ❖

Physicists in the mid-1930s ignored far more than Hitler; they also ignored nearly all of nuclear physics. They far preferred atomic or molecular physics, the structure of water molecules, or metallic strontium. Nuclear physics was hardly touched.

The first news of the Hahn–Strassmann discovery came to the United States in January 1939, with Niels Bohr, who was visiting Princeton for six months. I was then laid up in the hospital for about six weeks with jaundice. Leo Szilard was staying in my apartment.

Jaundice is a condition most often caused by a disease of the liver or biliary tract. The bilirubin in your tissues makes you look yellow and queer, and you feel powerless. I was at first depressed and drowsy, unable to think clearly about physics or anything else. My diet was nothing but potatoes and beans, boiled in water.

Szilard came by the infirmary nearly every day to see me and raised my spirits with gentle Hungarian conversation. I appreciated that tremendously. Jaundice does not really hurt, and soon I felt quite at ease. My recovery gave me time to rest, and I felt a striking detachment from the daily cares of scholarship.

One morning, Szilard came to my bed and said, "Wigner, now I think there will be a chain reaction." He meant that a nuclear fission would be possible. I disagreed at first, but soon I saw that of course he was right. I had miscalculated how neutron-rich were the fission fragments.

Szilard and I intently discussed uranium fission and the likely effect of the Hahn–Strassmann finding. We already understood chemical reactions, so we foresaw the general shape of fission theory. In the course of these talks in the Princeton infirmary, Szilard and I developed all of the essential points of fission theory.

In late January 1939, a large theoretical physics conference convened in Washington, D.C. It had been organized by George Gamow. I was unable to attend because of the jaundice, but many of the best physicists in the country were there and nearly everyone was talking about nuclear chain reactions. The prestige of this group confirmed the importance of fission. Enrico Fermi made a formal report on the discovery of fission and the practical consequences likely to follow. Szilard made a clear, careful memo on the conference and gave me a copy.

Soon afterward, Niels Bohr and the Princeton physicist John Wheeler independently published their own superb papers on fission theory. Both men were near-geniuses. Bohr often spoke in generalities. His scientific style was intuitive, at times obscure, but vital and charming. Reading the work of Bohr and Wheeler, I was pleased to see that in comparing the stabilities of uranium nuclei, Szilard and I had seen farther than those two at several points.

Szilard and I discussed all this with Enrico Fermi. Szilard correctly predicted that neutrons would be emitted during the fission process, a result that Bohr and Wheeler doubted. But neither Szilard nor I saw that the slow neutron fission was due only to the uranium 235 isotope. That vital insight belonged to Niels Bohr alone; Szilard and I were skeptical of it until it was verified experimentally.

Fermi foresaw much of this, but said little. He was a far more cautious predictor than Bohr or Szilard. Another Princeton man, Louis Turner, predicted both the existence of plutonium and the fact that it would be fissionable. And Henry Smyth, who had been teaching in the Princeton physics department for 15 years, was also quite capable in these matters.

So, as a Princeton professor, I was well placed to follow advances in fission. The Princeton Physics Department had been slow to grasp quantum mechanics; happily, they saw quickly the importance of fission and soon we were conducting fission experiments.

Richard Feynman was another Princeton scientist who helped us defeat Hitler. He was a young graduate student in 1939, with a remarkable brain—almost another Dirac, but far more emotive. Feynman gave a talk in 1939 to which I invited the great astronomer Henry Norris Russell, Johnny von Neumann, the eminent physicist Wolfgang Pauli, and Albert Einstein.

Feynman felt awed at first by his audience, as I had once felt at the University of Berlin colloquia. Albert Einstein often had that effect on people. But just as I had learned in Berlin, Feynman soon found that Einstein was not at all intimidating if you knew your subject. Feynman, too, was an expert in fission within a few years. There were many talented people at Princeton studying questions closely related to nuclear fission.

In trying to create an atomic bomb, the novel question was: Could we maintain a very large chain reaction? In chemistry and nonnuclear physics, no such enormous chain reaction exists. And the scale of everything in building nuclear weapons is enormous, which required a mental adjustment.

We worked partly with the theory, partly with laboratory experiments. Theoretical physicists work mostly at home. But I persuaded some clever young experimental physicists in Princeton to work on the atomic bomb. While investigating resonance absorption, I convinced a wonderful young physicist named Robert Wilson to help me. Wilson later became head of the National Accelerator Laboratory.

And I persuaded Edward Creutz to study the nuclear physics of uranium. Creutz was later a top director of the National Science Foundation. Wilson and Creutz were strong, skillful men, who worked with a vivid understanding. We conferred constantly with Enrico Fermi's group at Columbia.

And we found something quite useful: Whether the resonance absorption stopped the chain reaction depended on the size of the uranium lump. We got little, if any, money for all this from the government. But neither did we have to worry about government meddling or red tape. I was still working as a Princeton professor only. I alone chose the subject of my research. I must have told a few close Princeton associates about it; no formal government security yet forbade such talk.

Szilard tried valiantly to enhance fission knowledge on the one hand and to keep it secret on the other. By April 1939, two distinct ways of getting fission seemed promising, though each had drawbacks. A chain reaction with slow neutrons seemed possible, but for power production only. A fast neutron process looked feasible, but only for violent explosions. Niels Bohr and John Wheeler felt that the chain reaction could not be achieved, except perhaps at low temperatures.

Bohr also said that once Fermi had spoken publicly of fission at the Washington conference, it could no longer be kept secret. Szilard, Edward Teller, Victor Weisskopf, and I got Bohr to agree to delay publishing on fission if the British and Frédéric Joliot's French group would do the same. Letters and telegrams were dispatched to Joliot, Hans von Halban, Patrick Blackett, and Paul Dirac with this idea.

The British agreed, but Joliot would not. I do not know why; perhaps he feared that a failure to publish would cost him some fame or hurt his reputation. And his fears were partly justified; some sensational reports did appear in American newspapers.

At any rate, Joliot killed our hopes of delaying publication. We were annoyed with him, but what could we do? And once Joliot had published, we dropped the request. Szilard, Fermi, Hans von Halban, and Lew Kowarski followed with similar work in *Physical Review.* I had some views of my own, but did not publish them, partly to avoid competing with Leo Szilard.

I played the interested spectator.

∾ *Twelve* ∾

# *Swimming in Syrup*

*E* agerly, Leo Szilard and I discussed the atomic bomb with our university colleagues. Szilard knew the crucial questions: How could we reach the men who controlled American warfare? And how could we persuade them of the military value of an atomic bomb project?

Our academic colleagues, for all of their eminence, had little authority in warfare and few resources. Szilard knew that we could hardly make an atomic bomb without federal support. But he disliked this idea; and like most men he was reluctant to face that which he disliked. He was afraid that the government would saddle uranium research with an awful load of bureaucracy.

I insisted that the Roosevelt government be kept informed of progress in the nuclear chain reaction. Others whom Szilard respected did, too. Szilard fussed and protested, but inside I think he agreed. He saw that we had floundered on the publication question precisely because we could not speak with authority for all of American physics. Bringing in the federal government could give us that authority.

Szilard and I asked each other: How should we contact the government? We knew a dean at Columbia University named George Pegram, who in turn knew many higher-ups in the federal government. We asked George Pegram for help.

Pegram telephoned an under secretary of the Navy in Washington named Edison. After that call, Enrico Fermi was invited to meet some Navy men, including a man named Ross Gunn. The talk was a distinct failure. Fermi got nothing more from it but 2000 dollars for isotope research to be conducted by the physicist Jesse Beams.

You see, the Army and the Navy had their own thinking, and they found atomic bombs alien. To Szilard and Wigner, atomic bombs already had a certain reality. But what did the Army and Navy see? They saw a few European refugee scientists pleading for a project that was almost entirely a fantasy.

So the Army and Navy did the natural thing. They said, "Let's ignore all of this. It's just some crazy worrying by a few foreigners." And, in a way, they were right. There *was* something crazy about touting an atomic bomb, when so little of the crucial work had been done.

Szilard and I decided to enlist Albert Einstein's help. We wanted to write the Belgian government, to warn them to keep the uranium ore of the Belgian Congo out of reach of the Nazis. Einstein somehow knew the Belgian Royal Family, so we felt he ought to write this letter. It was not customary for physicists to be on familiar terms with European royalty, but then Einstein was not an ordinary physicist.

We feared that it might be illegal to initiate private conversation with a foreign government on a potentially military matter, especially since Szilard and I were not "real Americans." So I suggested that we send a copy of Einstein's letter to the State Department, with a note attached warning that the letter would be posted in two weeks time unless we were told to hold it.

Szilard wanted Einstein to write a letter to President Roosevelt, too. Teller and I agreed with Szilard. We called up Einstein's house in Princeton and were told that he was off in Peconic, Long Island.

July 16, 1939 was a Sunday. Szilard and I drove out to Peconic and got terribly lost trying to find the house. We knew the street name, and along the way asked a good many people where the street was. No one could tell us. Finally, we saw a young man walking, and called over to him, "Say, do you know where Einstein lives?" And he told us very directly. We kept to his directions and reached the house. There we found Einstein dressed in an old shirt and unpressed pants, apparently perfectly content to be thinking only of physics.

You should know that Einstein was not interested in nuclear physics. Indeed, he thought that all of quantum theory was largely an error. And who knows, perhaps someday Einstein will be proved right on that; though, if so, it has certainly been quite a useful error.

I believe I had already spoken with Einstein a few weeks before about a possible nuclear chain reaction, but only lightly. On this July day, we spoke deeply. As usual, we spoke German. Einstein was quite receptive. I thought it would take time to persuade him to write President Roosevelt. But no; though he had rarely thought about nuclear chain reactions before, within ten minutes he grasped the situation. He understood the meaning of a chain reaction scientifically, remarking that for the first time in history, men would tap energy from a source other than the sun.

Einstein also realized the political and military meaning of nuclear fission: that it could yield explosives strong enough to make the Nazis invincible. And Einstein was just as horrified as I was by that prospect. He volunteered to do whatever he could to prevent it.

Let me sketch Einstein's politics in 1939; they have often been misunderstood. Einstein was still a year away from becoming a naturalized American citizen, but already he admired the American political system. He was interested in the destiny of the United States. At least one Einstein biographer has written that Einstein hated the German people. That is a sad distortion. At times, Einstein spoke imprecisely about politics, but I cannot believe that he ever felt hatred for the German people.

What Einstein detested was the Nazi regime; and he hated it for killing political and intellectual freedom as well as for attacking the Jews. Einstein had bitter words for dictators of the left as well; he was angry with Russian communism as early as 1932. Einstein was a man with very strong convictions about what is good and what is evil. But he respected the German people.

Like many young men, Einstein had once felt an aversion to all things military. But by 1939, he had seen that unwavering pacifism does not contain a bully. Einstein had the future of freedom very much in mind on the day we found him in Peconic, and he knew the value of a strong military.

I knew that Einstein would agree with us about the danger posed by Adolf Hitler. That he would agree to dictate a letter to President Roosevelt, I was far less sure. Einstein was never an overtly political man, and he preferred to stay in the background when politics was argued.

But Einstein quickly agreed to dictate a letter to the president. And he dictated it immediately. It was uncanny how readily Einstein did this. To formulate an important series of ideas quickly and clearly is not easy, but Einstein did it, and his letter contained not one error.

Einstein spoke in German and I scribbled down his words. He spoke of uranium and warned of the possibility that a nuclear chain reaction could be harnessed to produce a terrible bomb. He called for watchfulness by the White House and, if needed, strong action.

Back in Princeton, I translated the letter into English. So I am largely responsible for the English wording of the famous Einstein letter. But the translation was simple because Einstein's German was so precise.

The letter went back to Einstein and he signed it without hesitation. After some final revision, the letter was dated August 2 and sent to President Roosevelt.

Within two months, England and Germany were at war.

A man named Alexander Sachs brought the Einstein letter to President Roosevelt in Washington. It helped start the Uranium Project. For the rest of the war, I never spoke to Einstein about my work on the atomic bomb. He must have generally known of it; he followed the American war effort, at his usual distance. Einstein was never formally asked to join the Manhattan Project. Most of the work on the bomb was quite specialized and closer to engineering than to physics. This did not suit Einstein. And apparently Einstein's history of pacifist statements had irritated the military higher-ups. They did not find Albert Einstein a suitable scientific warrior.

Einstein actually did a bit of minor work for the war effort. A man of his stature is constantly approached by outsiders. At one point, the U.S. Navy told him, "You are Einstein. It cannot be hard for you to make a few theoretical calculations for us." They were right; it was not hard for him and he did the work. But I hardly tried to speak with Einstein about our wartime work, knowing that it was not allowed. Secrecy became a habit in that war.

I never communicated directly with Franklin Roosevelt or with Secretary of War Henry Stimson. So Szilard and I were pleased and a bit surprised when we heard that a meeting had been set up at the Bureau of Standards to discuss the atomic bomb.

Gregory Breit, my wild but loyal friend from the University of Wisconsin, was working on national defense for the Bureau of Standards. Breit planned this meeting, but we felt that the meeting would not have occurred unless some top men in the federal government, perhaps even including President Roosevelt himself, were finally heeding our warnings.

President Roosevelt was in no way a scientist. But if he did not believe that nuclear chain reactions already threatened world peace, he may have seen the future threat. He wanted a small committee to link his government with the thoughts of top physicists.

❧      ❧      ❧

That summer of 1939 brought some crucial advances in physics. First, Fermi and Szilard decided to search for the slow neutron chain. Second, they abandoned hydrogen as the slowing-down substance because it had too high a neutron absorption. They chose graphite instead. Third, they proposed a latticelike arrangement for the uranium and graphite rather than a uniform mixture. This idea was more likely Fermi's than Szilard's.

The first meeting I attended at the National Bureau of Standards was in Washington, D.C. on October 21, 1939. Lyman Briggs, chairman of the so-called Uranium Committee, was there. An Army colonel named Adamson was there and a Navy commander named Hoover. A man named Meggers represented the Bureau of Standards, and two other men, Roberts and Abelson, were with him. Two local physicists were there, Fred Mohler of the Bureau of Standards and Richard Roberts of the Carnegie Institution.

Alexander Sachs was there, too, the man who had helped put Einstein's letter into the hands of President Roosevelt. Sachs had also invited Einstein, Leo Szilard, and Edward Teller. Teller had been asked partly because he was teaching at nearby George Washington University and could keep a better eye on things in Washington than his friends Wigner and Szilard. All of those invited had come—all except Einstein.

Early in the meeting, Szilard pointed out that making a chain reaction would require money to perform experiments and to acquire uranium supplies and graphite. Szilard also suggested creating a permanent committee between the physicists and the government.

Teller and I supported Szilard. But Roberts and Abelson opposed him. The colonels kept rather aloof. I had the feeling that these soldiers and government men were just like the others I had approached: they were friendly, they smiled, but they never expected to see a working atomic bomb in this world.

Teller again raised the need for funding. One of the military men asked how much we would need. Szilard suggested 6000 dollars. I never knew how he reached that figure, but there it was. There was a little unhappy pause from the military men, and then Colonel Adamson said, "Gentlemen, armaments are not what decides war and makes history. Wars are won through the morale of the civilian population."

Well, Teller, Szilard, and I did not believe that statement at all. I doubt Alexander Sachs did either. I got a bit angry and spoke up, almost for the first time: "If that is true," I said, "then perhaps we should cut the Army budget 30 percent and spread that wonderful morale through the civilian population."

Colonel Adamson flushed. Apparently, he did not consider the dismantling of any military service a good idea. He slapped the table angrily with his hand and said, "How much money do you say you need?" We got our 6000 dollars.

I tell that story at the expense of Colonel Adamson, but in a way he was right. As an old man, I see his point much better; the morale of a nation often *is* decisive in a war. And some years later, I decided that the Manhattan Project really began at that meeting; that the Bureau of Standards, however inactive they were in the Manhattan Project, had yet stood in the background, steadily pushing things forward.

But to a headstrong young group of Hungarians, that was not clear in 1939. At that time I was irked by this meeting at the Bureau of Standards—dismayed by the military men who ran the meeting and by the government that supported them.

The years from 1939 to 1941 made an alarming impression. The forces of democracy seemed to be sleeping, while the fascist states were aggressive and alert. Germany seized for itself the Norwegian "heavy water" useful in making atomic bombs. The Germans marched into Czechoslovakia and plundered their mines, among the greatest in Europe. And they stopped the Czechs from selling uranium internationally.

I knew firsthand a number of German scientists who knew enough to create a crude atomic bomb. Otto Hahn, who had first split the atom, was just one. I had a terrible fear that our government might never try to build an atomic bomb, while the Germans perfected one.

On April 27, 1940, another meeting of the Advisory Committee on Uranium was held at the National Bureau of Standards. The same people attended, and again Leo Szilard and Alexander Sachs argued that work on a nuclear chain reaction should be accelerated. I quietly supported their views.

But getting the U.S. government to see the value of fission felt like swimming in syrup. We learned that governments do not like fantastic new projects. And leaders can hardly be blamed for that. In wartime they are assailed by hundreds of schemes for "miracle" weapons. And no scientist could honestly say that an atomic bomb was sure to work.

No more decisive uranium work was done, and Szilard and I got quite nervous. I often urged Szilard to publish his views, confronting America's leaders with the need for action. We both had to push very hard to get the attention of our colleagues.

Precious months passed, and our exasperation grew. I felt that our government, consciously or not, was conceding to our enemies and abandoning our future allies; that our country was weakening both materially and morally. But none of the scientists working unofficially on the atomic bomb could really hope to change things. The authorities thought us dreamers enough as it was.

In this uncertain time, we all owe a debt to Gregory Breit, Harold Urey, and many others who hastened matters. Breit was a natural hastener. His determination helped produce a Reference Committee that judged fission-related papers to see if they should be published in scientific journals. And in this role Breit the hastener was paradoxically quite cautious.

Around 1940, I met a woman who would change my life: Mary Annette Wheeler. In later years, she and I could never agree on just where we met. But I think it was at a summer school of theoretical physics held at the University of Michigan in Ann Arbor.

I could never describe all that Mary Wheeler meant to me, nor should I. But I ought to give a quick sketch. She had grown up in New England, had studied at Wellesley and Yale, and was now teaching physics at Vassar. I was immediately intrigued that she was a New Englander. But that she was also a physics professor—that was even better. There were very few female physics professors then, even fewer than there are today.

People sometimes ask, "Why are there so few female physicists?" Well, women are hard to know. They have their own traits. Yet many women have been fine scientists despite the familiar diversions of bearing and raising children. So I prefer not to ask, "Why do so few women embrace physics?" but rather, "Why do so many men embrace physics?" That is the far more remarkable thing.

Mary Wheeler was an emotional woman, honest and upright. I soon realized that I responded to her ideas and emotions in a quite unusual way. She was easy to talk with, and we began talking a great deal. She liked walking, too, so we took walks together.

And unlike most women, Mary Wheeler liked to talk about physics on her walks. I cannot say how much this delighted me. She also told me of her family, her past, and her desires. It was perhaps on a walk near the woods of the Princeton University Graduate College that I asked Mary Wheeler to marry me. I felt that we now knew each other well and should not loiter around. We should marry. To my great pleasure, Mary Wheeler agreed. On June 4, 1941, she became Mary Wheeler Wigner.

Marriage came naturally to us both. We were nearly the same age and had a great natural fondness for each other. I cannot say what Mary most liked in me. But she knew it was time to marry, to spend her life with a loving and honorable man. And both of us wished to pass along our physical properties: the qualities of our brains, and even of our bodies. Before long, Mrs. Mary Wheeler Wigner gave me two precious children, David Wigner, and then two years later, Martha Wigner. They grew up to resemble us quite nicely.

Having lived some years as a bachelor, I already knew something about keeping a household. My wife was never a scold and I was not a demanding husband. We went to films and theater together, went ice skating, and shared an emotional life. Marriage is wonderful; I do not understand why so many adults are unmarried.

By 1941, the happiness of my marriage and the growing national interest in resisting Hitler had largely eased my deep gloom of 1938. But I remained apprehensive.

In mid-September 1941, the eminent American physicist Arthur Compton turned up in Princeton to see me and to gauge the work of my physics team. Arthur Compton was a brother of Karl Compton, the man who had spent six months at Princeton without learning the difference between Johnny von Neumann and Eugene Wigner.

Arthur Compton would not likely have made the same mistake. He was a large man with a dark mustache, quite handsome without trying to be. With monochromatic x rays and a crystal spectrometer, Compton had measured the change in wavelength of x rays scattered at various angles from a single target. He had done this beautifully in the mid-1920s, and soon we were all calling it the "Compton effect." It earned him a Nobel prize in 1927.

The Compton effect had roused Berlin. I was then doing my engineering thesis under Herman Mark. At first Mark had said, "Bah. Listen to me. No such Compton effect exists." But finally, even he was convinced. And my fluctuation calculations under Mark were more derived from the Compton effect than from anything else. So I had great respect for Arthur Compton as a physicist.

Now, in September 1941, I told Compton the difference between fast neutrons and slow neutrons; and that a nuclear reactor with enough uranium and a strong moderator could well make the needed plutonium for an atomic bomb. I argued with feeling that Compton had a duty to start an atomic bomb program and to speed it along as best he could.

Compton later wrote a book called *Atomic Quest,* a fine account of his work on the Manhattan Project. An apparently vivid part of his atomic quest was the day that Wigner urged him, "almost in tears," to help our nation build the atomic bomb. Well, I certainly did not sob as I addressed Arthur Compton that day. But he should not have been surprised that the thought of Nazis deeply stirred my emotions.

On December 8, 1941, the very day after the Japanese bombed Pearl Harbor and brought America directly into the war, I heard from Arthur Compton again. He told me that the Uranium Project was being reorganized and that he was heading all official study of graphite–uranium lattice reactors. This work had begun some months before at Columbia University in New York City, but Arthur Compton was well settled at the University of Chicago and he wanted the chain reactor program located there.

Compton said that the U.S. government would be starting a special war laboratory in Chicago and transferring the Columbia and Princeton University groups there. He offered me a high position in the new Chicago laboratory. He told me that he was also inviting Enrico Fermi and other fine scientists, many of us foreigners.

Most Americans in 1941 felt that recent immigrants to

the country should not get secret war work. So when the first department was started in Cambridge, Massachusetts, working mostly on radar, it was filled with Americans of old standing. I did not mind. I understood why the United States disliked sharing war secrets with "foreigners."

But I was relieved to find a laboratory for new Americans like Fermi and Wigner. Compton told me very little about my future duties. He did not have to. It was enough to know that we would try to create a nuclear chain reaction. I had waited years to hear that from a government official.

I stayed at Princeton for about three months more, while my Princeton group studied the theory of nuclear chain reaction. Then in April 1942, I took a formal leave of absence from Princeton University and my wife and I went to Chicago, where I joined the Metallurgical Laboratory.

I was not supposed to tell my wife about the atomic bomb. But it was impossible not to. You cannot easily hide your goals and dreams from those you love. And I saw no reason to try. My wife knew much of the relevant physics already. She was a real American and utterly trustworthy by nature. So I told her everything.

We closely discussed the meaning of the work in Chicago before deciding to move there. But I always knew that I should go. I was not a leader by temperament and I took a leader's role reluctantly. But now I felt it was imperative.

For many months, I had been pushing for a federal program to make atomic bombs. Now that one was forming, I felt bound to join. I was quite conscious of an immoral element in my action. But I was far more concerned with the moral failings of a man across the ocean: Adolf Hitler.

∾ *Thirteen* ∾

# *Martians*

*I* am not a great man and have never had illusions of greatness. I would much prefer to be a soldier than an officer. The job I took in April of 1942 suited me well. I was the head of a group at the Metallurgical Laboratory on the campus of the University of Chicago. I served a useful purpose, but in the large scheme of the Manhattan Project, I was, so to say, a soldier, not an officer.

Just as the "sanitarium" where I stayed as a child was not chiefly concerned with the sanity of its patients, so this Metallurgical Laboratory was not designed for metallurgy. It was not until late in 1942, when we realized how little was known about the metallurgy of uranium, that the Metallurgical Laboratory truly became a metallurgical laboratory.

The name disguised a laboratory that was trying to create a nuclear chain reaction. And not any nuclear chain reaction, but one of enormous power, which is quite a different thing. A tiny chain reaction shows that chain reactions are possible;

but to make and control a huge chain reaction, you must extract its heat, measure its power, and perform many other irksome tasks with great precision.

On December 7, 1941, the Japanese had attacked us at Pearl Harbor and formally declared war. This attack had finally aroused the American people and united them for war. After Pearl Harbor, money and manpower came to us in an ever-growing stream. The war that followed allied the United States with Russia, Great Britain, France, Belgium, Poland, the Netherlands, Czechoslovakia, Greece, Norway, Denmark, and Yugoslavia. We fought the forces of Germany, Italy, Japan, Austria, Rumania, Bulgaria, Finland, and, sadly, my beloved Hungary.

I recall Enrico Fermi saying, "From now on, our problem will not be too little money, but too much." Fermi knew that government money brings bureaucracy. And so it did. Over time we had to teach hundreds of people the art of a nuclear chain reaction. The teaching delayed the bomb's production by about nine months. But without the Army, an atomic bomb could hardly have been built.

The U.S. Army ran the Manhattan Project with a great devotion to secrecy. The name "Manhattan Project" was part of the disguise. The name grew naturally from the Army's Manhattan Engineer District, first located in Manhattan, New York. But the name carefully avoided describing any atomic bomb.

I often wondered precisely how much the Army enjoyed secrecy. They were certainly well trained in it. They had precious ideas about our military defense and claimed that national security demanded that these ideas be kept secret. Also, I think the Army feared that a working atomic bomb might prove impossible to build, and they did not want to be embarrassed if it could not. So the Army split the Manhattan

Project into many smaller projects, each with its own private veil of secrecy.

The Met Lab in Chicago was one of these secret army stations; there were others in Hanford, Washington, and Oak Ridge, Tennessee. Perhaps the most important was in Los Alamos, New Mexico.

General Leslie Groves was the head Army officer for the Manhattan Project. What an interesting man Groves was! I must admit that many of my physicist colleagues quite disliked him. But I liked Groves. He was a boss in every sense of the word. He had the temperament of a boss and held the assumptions of a boss. But in most ways he was a likeable man.

General Groves had the body of a boss, too, really a quite striking physique. He was more than a little fat; and he must have known that he was. When you are as fat as General Groves, you are aware of your girth. But fatness does not discourage a general. Groves was a dynamic leader who used his great brawn to effect.

General Groves was charged with building a large, complex organization quickly and well. He worked enormously hard during the war. He first came to the Met Lab in October 1942, talked to the scientists, looked over our equipment, and asked us how much plutonium and uranium 235 we would need to make a working atomic bomb. From then on, General Groves frequently spoke with the scientists, praising and supporting our work. Sometimes he criticized our work and even our temperaments and backgrounds, but most of his concerns were reasonable.

General Groves was not a scientist. He did not know physics. Apparently, he pretended some scientific knowledge, especially in the spring of 1943 when the serious work at Los Alamos was beginning and Groves went out to deliver some pep talks. Perhaps General Groves should not have pretended to know physics. His scientific pretensions annoyed some physicists.

But I am sure that deep down General Groves knew how ignorant he was of science. And he knew that we knew it too. General Groves never tried to practice science. He acted the general's role, leaving to us the details, the precise scientific calculations.

I wish I could have known General Groves better. I would have liked to take leisurely walks with him, to tell jokes, discuss broad moral issues, and share extended meals. I would have liked to watch him eat. But it was wartime, and such luxuries could not be afforded. Certainly not by General Groves. I contented myself with a more distant view of the boss.

My wife and I rented three rooms near the University of Chicago. Our first child, David, was born in 1943. During the day, my wife spent much of her time caring for our baby. She also worked for the university, though not for the Uranium Project.

The Met Lab occupied most of Eckart Hall, a large building that housed the University of Chicago physics and mathematics departments. The Met Lab also took up much of Ryerson Hall and some of the north and west stands of Stagg Field.

But I focused on the work of about 20 physicists on the fourth floor of Eckart Hall. I had no formal title, but I led a theoretical group of about seven. Most were young theoretical physicists with little background in experimental physics or engineering. We worked in two or three spacious rooms. One room was my office. I asked people in to discuss technical problems.

Two methods of fission were then known. The first was based on uranium 235, the rare isotope of uranium. The second was based on plutonium 239, a new element that could be produced in a nuclear reactor.

The chief task of my group was to design a nuclear reactor that was to be used in the Manhattan Project station in Hanford, Washington. We felt sure that a nuclear chain reaction could be made and controlled. We even knew crudely what had to be done: to make a chain reaction producing plutonium, and to understand the separation of the plutonium and the dissipation of heat that the fission process produces. We even knew two ways to approach the chain reaction.

So the mood on the fourth floor of Eckart Hall was pleasant, almost familylike. My hours were fairly short: I began about 9:00, often lunched at home, and usually quit by 6:30. The few times I missed work, the others worked just as well.

I rarely gave my staff personal advice. I gave them problems to calculate and gauged their work with a critical eye, suggesting to them how to avoid potential troubles. Working with these young men was a pleasure.

But there was another side to the work, one far less pleasant. There was pressure to foresee every trouble, a constant fear that with one small error the Germans might beat us to the atomic bomb. I felt this fiercely.

An American security force wanted my fingerprints. I refused to give them. A fingerprint record might someday fall into the hands of the Nazis. I had no doubt that if the Germans won the war they would swiftly begin rounding up everyone in the Manhattan Project for execution. And the rounding up would go easier with fingerprints.

Thoughts of being murdered focus your mind wonderfully. So there was a warmth in the Met Lab, but also a fear: a warmth of human relations and a fear of losing the race. Nineteen forty-three was the first year in twenty that I did not publish a single scientific paper. Very little was published during the war; we were too busy.

Fortunately, most of us did not allow the hard work and the fear to warp our collegial relations. We learned more about human nature than we had in all the previous years of academic seclusion. But considering the stakes, the human friction in Chicago was quite small. There were major disputes on technical points; but then good physicists should never fully agree on any new piece of physics. I believe the atomic station at Los Alamos had more friction than we did.

All of my young collaborators were superb. Francis Friedman, Gilbert Plass, Lee Ohlinger, Marvin Goldberger, and Frederick Seitz were especially fine. Today, some of those names are obscure even to American physicists. Fifty years ago, they were all young patriots.

But my two favorite collaborators were Alvin Weinberg and Gale Young. Weinberg and Young shared the room next to mine. Both were able, farsighted, and deeply likeable young men. We never quarreled. I gave them serious advice and also joked with them like a good Hungarian. We became strong partners and friends. I was a teacher to Alvin Weinberg and Gale Young almost as Polanyi had been my teacher. And that pleased me deeply. But Weinberg and Young were excellent pupils. They would have thrived without me.

Alvin Weinberg was American-born, short, trim, and dark. He was a biophysicist by training who swiftly taught himself a great deal of the physics of nuclear energy. Weinberg beautifully understood human nature and the course of human relations. He knew that there is such a thing as ambition and also that there are many ways to accommodate it. He sized men up quickly and was pleasantly persuasive, a natural diplomat.

And you are very lucky to find a natural diplomat who

also understands nuclear energy. We gave Alvin Weinberg a great many nontechnical tasks. We set him working on broader scientific questions. His grasp of human personality won over many doubters. Weinberg never failed us.

Gale Young was also American-born, short, wide, and handsome. Before joining the Met Lab in March 1942, he had headed the mathematics and physics departments at Olivet College in Michigan. That experience seemed to have taught him how organizations are run and how they ought to be run. He was not a great physicist and did not pretend to be. But he had the devotion and the mental outlook of a superb engineer. Others knew more theory, but Gale Young always knew the practical applications.

Young wanted to be the best. When he made a mistake, he saw it quickly and continued, undismayed. Determination is a quite underrated quality; by force of will, Gale Young became our best theoretical engineer, with a special knack for knowing where to make adjustments in construction.

So it was natural that I made Gale Young my main helper in 1942 in designing a nuclear pile cooled by water. By then we knew that natural uranium would be our reactor fuel and graphite our moderator. But a nuclear pile quickly becomes terribly hot and must be cooled. What should the reactor's coolant be?

Four competing methods were conceived: so-called "heavy" water, bismuth, helium, and air. But I was convinced that by the time all the engineering troubles involved with helium were resolved, Adolf Hitler would have conquered much of the earth. So I urged the adoption of a proven cooling method: ordinary water. Gale Young and Alvin Weinberg backed me up.

Some of the engineers felt, with some justice, that we physicists were treading on their territory. So the choice of coolant was complex not only scientifically but politically within the tight circle of the Metallurigcal Lab. Some of the engineers gave my group the friendly label, "The Fourth Floor Communists."

Gale Young and I hardly noticed. We were busy drawing up the design of a water-cooled pile. We worked too swiftly to produce anything elegant. But we showed this preliminary design to Arthur Compton around June 1942. And though Compton had been one of the chief backers of helium cooling, he liked our design and urged us to improve it.

By July 1942, Gale Young and I had rough plans for a great 100,000-kilowatt pile to be made of uranium lumps in a graphite cylinder 12 feet high and 25 feet around. It would be cooled by water. This design was further improved throughout the year, and eventually became the basis of the nuclear reactor built in Hanford, Washington.

The design that my Met Lab group finally adopted for the nuclear reactor was a roughly cubic form, about 6 yards in each direction. We figured how to insert the uranium, how to remove it, and many other such things. It seemed as complex as designing a factory. I think we did it well.

Making a giant chain reaction work began with physics. But it soon required engineering. Before you can start production, you must overcome a great many small engineering flaws. And we had few design engineers in the Manhattan Project then.

So the theoretical physicists had to calculate the dimensions of an I-beam. We had to know how aluminum tubes are fabricated and corroded and how to calculate water flow rates in these tubes. We had to know that radioactivity is induced in oxygen by neutron absorption and that the damage to the structure of graphite decreases with its thermal conductivity.

So I was glad of my engineering training. Scientists who had begun in self-contained theoretical and experimental groups were mixed up together and came to know each other well.

My friend Edward Teller also worked at the Met Lab. He had studied at Göttingen from 1931 to 1933, under James Franck. Teller had served his mentor with unswerving loyalty. When Hitler came to power, Teller had gone to Copenhagen to spend a year studying with the great Niels Bohr.

Teller had lectured at the University of London. In 1934, an old Copenhagen colleague, George Gamow, had come to the United States as a physics professor at George Washington University, in Washington, D.C. Gamow was the same man who later tried to help me and Szilard reach the President with our fears of atomic bombs. In 1935, Gamow convinced George Washington University to extend professorships to my friends Edward Teller and Gregory Breit. So Teller had reached the United States with Breit in 1935.

Between 1930 and 1936, Teller had written more than 30 papers on subjects ranging from cosmology to atomic theory. They were extraordinary works, illuminating, erudite, and ingenious.

During 1941–1942, Teller had taught at Columbia University. But he had abruptly given up his study of pure science to join the Manhattan Project. There he had greatly improved the theory of nuclear chain reactions. He disliked the usual global description of neutron population. In order to trace individual neutrons from birth to absorption, he created a deeply insightful theory.

Perhaps Jancsi von Neumann had a brain more powerful than Teller's; and certainly Einstein created more momentous physical theories. But Teller's imagination was more fertile than that of anyone else I have ever known. He did not linger over elegant mathematical formulations, like most theoretical physicists. He studied the phenomena themselves, using available, empirical data with brilliant insight.

Jancsi von Neumann worked for the Manhattan Project too, though only in a general way. He was a consultant to both the Army and Navy. They would ask his opinion on various questions, saving him the hard ones. Jancsi would give his opinion, to which they would respond, "Very fine, Johnny. You know, you are really remarkably intelligent." Jancsi would grin. He knew very well that he was a genius; he had been hearing it since he was a 10-year-old boy studying higher mathematics in Budapest.

"Implosion" means a violent collapse inward, and implosion was a means of igniting the atomic bomb. Many people helped develop the theory of implosion, among them George Kistiakowsky, Seth Neddermeyer, and Robert Christy. But Jancsi von Neumann did as much as anyone to conceive the idea. Jancsi had a surprising knowledge of explosions, and his precise calculations helped convince Robert Oppenheimer and others that implosion would properly ignite the atomic bomb.

I never knew why Jancsi was not asked to play a more central role in the Manhattan Project. He would have found the project fascinating. Apparently, some of the top leaders felt that he had been away from these things too long to be effective. Perhaps they thought him too absentminded; people often assume that all geniuses are absentminded. It is a foolish assumption.

Once I asked Arthur Compton, "Say, why don't we bring von Neumann more closely into the project?" Compton enjoyed the suggestion, but he said that it was too late; we were too far along and Jancsi would have to master too much new material. I have no doubt that von Neumann could have caught up with us quite easily, and done far more for the Manhattan Project than he did. But he was never asked.

✻ ✻ ✻

The Hungarian scientists working on the Manhattan Project—principally Szilard, Teller, von Neumann, and Wigner—were thought to be queer by the Americans. They often called us the "Martians." The label was unreasonable. We Hungarians had no more contact with Mars than they did. But we had imagination and some far-reaching desires. The planet Mars is quite distant and was then quite prominent in the mind of American science. So the Hungarians were often accused of having come from Mars. Werner Heisenberg, Enrico Fermi, and Niels Bohr also had ideas of extraordinary reach. And yet Germans, Italians, and Danes were never called "Martians."

Well, perhaps we Hungarians kept a bit to ourselves. That was natural. We came from a special nation, with its own distinct history, language, and ways of thought. Others found us a bit strange.

That was all right; we found Americans a bit strange. Hungarians were raised to be far more respectful in their contact with ladies. We found curious the degree of intimacy between unmarried Americans, though I hope we were polite enough to keep quiet about it.

I think I was the only Hungarian scientist who wished to be a normal American. Szilard, Teller, and von Neumann liked being called "Martians." But I did not.

Many people have asked me: Why was this generation of Jewish Hungarians so brilliant? Let me begin by making clear it was not a matter of genetic superiority. Let us leave such ideas to Adolf Hitler. How could anyone feel "genetically superior" to Enrico Fermi or Werner Heisenberg? Much of the credit belongs to the superb high schools in Budapest, which gave us a wonderful start. But a greater spur to our success was probably the fact of our forced emigration.

Emigration can certainly be painful, but a young man with talent finds it stimulating. Outside your own nation, you lack a ready place. You need great ingenuity and effort just to find a niche. Hard work and ingenuity become a habit. Often they are enough to earn you a place above natives of your adopted country quite as talented as you.

Perhaps that is why many of the greatest American physicists of this century have been émigrés, and not only European. Chen Ning Yang and Tsung-Dao Lee came from China to the United States as young physics students around 1950. In 1957, they jointly received the Nobel prize in physics. And other fine scientists have come to us from the Orient. Emigration stimulated them.

*         *         *

When I had first met Leo Szilard in Berlin, his eccentricity and selfishness had as yet found no definite purpose. But as Szilard aged, the purpose of his queerness seemed to become clearer. He wanted a high political position.

Szilard did not stop to consider the source of his desires. But I believe the root of his great ambition was a fear of being considered a man of no consequence. He seemed at times quite relaxed and content, at other times driven to pursue political influence. This very fear gave Szilard remarkable persuasive qualities.

Szilard had an excessive regard for his own talents. His thoughts revolved too much around himself and his own proper place in the world of affairs. And yet, though most conceited men are complacent, Szilard was incapable of complacency. He did not always see his own deficiencies, but he saw brilliantly many of the deficiencies of the world. And he worked very hard to correct them, often at some sacrifice to himself. Szilard was a true visionary, with all of the strengths and weaknesses of a great man. Yet theories about great men do not apply to Szilard, and general patterns of behavior cannot hope to describe him.

Leo Szilard was not one to hide his ambition. When he reached Chicago in 1942, he did not expect to be a common assistant. He said plainly and firmly that he deserved a high office, preferably full control of the Metallurgical Laboratory. Szilard had done important work in nuclear fission at Columbia University with Enrico Fermi, Walter Zinn, and Herbert Anderson. So he certainly deserved an important post at the Met Lab.

But I saw clearly then, and see even more clearly now, that Szilard certainly did *not* deserve the position of a boss. By 1942, he gave less love and attention to pure science than he once had. He was a lesser physicist.

It is never wise to seek prominence in a field whose routine chores do not interest you. I loved the daily work of physics, loved making physical calculations, even those that proved fruitless. Nearly every scientist in the Manhattan Project felt the same way. Szilard was an exception. He took no pleasure in extended calculation. And yet he refused to draw the logical conclusion: that he should not try to be a prominent physicist.

Moreover, Szilard did not understand human nature well enough to rule a vital group of scientists without listening to them. Though he was capable of listening, he never did it consistently. And Szilard scarcely grasped the value of organization and of meeting commitments promptly. So, you see, he would have made a poor wartime boss.

Szilard had original ideas but expected others to derive their implications. Now, it is true that when a physicist produces physical ideas that are clearly sensible and promotes them in the right circles, eventually other physicists will pick them up and develop them.

But many of Szilard's best ideas did not look clearly sensible. They looked crazy. And yet Szilard expected other scientists to work out these crazy ideas for him. That was unreasonable. When other scientists want to work on crazy ideas, they will work on their own.

So Szilard was not made head of the Met Lab. Around January of 1942 he was put in charge of the supply of materials for all the different Met Lab projects. He was supposed to be a kind of special assistant to Enrico Fermi. And Szilard certainly knew enough to serve Fermi very usefully.

Sadly, he did not. Szilard hatched many complex technical ideas in Chicago, both in physics and engineering. Like me, he tried to foresee potential problems and their solutions. He excelled at that. But few of his ideas were used.

Szilard was not a man who attracted protégés. What a contrast he was to Fermi, who disliked working alone and gathered a group of superb assistants around him. In Rome, Fermi had worked with Amaldi, D'Agostino, Pontecorvo, Rasetti, and Segrè. In New York, it was, among others, Herb Anderson, John Marshall, George Weil, and Walter Zinn. Fermi was a natural leader. They called him "the Pope."

By acting as a visionary, Szilard made some of the higher-ups feel that members of the Met Lab could not be trusted with routine administration. This is the strangeness of life, that what Szilard wanted most was something that poorly suited him. He wanted to rule.

In 1943, Szilard was naturalized as a United States citizen, and by then he aimed to be a political leader. If not president of the United States, then president of the Uranium Project or of something else important. A congressman at least; more likely a senator or minister. A dictator. If not direct power, he wanted the ancient recourse for ambitious Hungarian Jews: real influence over those with power.

Szilard liked making the small decisions of power. Most of us dislike being told, "You must attend a certain conference in Stockholm." And most of us dislike issuing such orders. But Szilard liked it. So Szilard ordered certain people around. Not me, happily, but others. He meant well, but he felt sure that since his opinions were correct, others should gladly take his instruction.

Szilard shared Plato's idea that society should be ruled by an elite. Szilard was never malicious; he had goodwill for all the stupid people. But he saw no reason for stupid people to craft national policy. Bright people should; people quite a bit like Leo Szilard. That some people are far brighter than others, that many levels of political influence exist in human society, these facts saddened and discomfited me. They delighted Szilard.

Since Szilard was a difficult man and General Groves an impatient one, I was not surprised when General Groves wished to fire Szilard. I may have interceded to prevent this; I hated to see Szilard or his work degraded. But I try not to remember the relation between Groves and Szilard. They were two men who rarely agreed, even on fundamental truths.

The earnest, cooperative pursuit of knowledge has a wonderful way of diverting forceful men from quarrels and power lust. But no matter how much Szilard learned, he remained quarrelsome and lonely for power.

Szilard was quite persuasive, but he liked to quarrel with von Neumann, and at those times I was hard-pressed to like and agree with both of them. Von Neumann was usually right; he was a better scientist than Szilard and a far more selective quarreler.

Teller, Wigner, von Neumann, Szilard—all of us loved science as boys and tried to influence politics as men. Teller, von Neumann, and I all became political conservatives. Szilard was nearly the opposite. Teller, von Neumann, and I tried mainly to boost the military strength of our adopted country, the United States. Szilard wanted to change the whole direction of American society and perhaps the direction of the whole world. He wanted to install himself as head of a new American government that would somehow enlighten the world about its greatest problems. Szilard had glorious general ideas. The details he never gave.

Szilard's admiration for Enrico Fermi was tinged with jealousy. During most of the time that Fermi was laying the foundation for the first large-scale chain reaction, Szilard could not bring himself to watch. He could not bear to play such an insignificant role. And yet as a scientist and an amateur historian, Szilard had a sincere interest in the proceedings. He asked me for reports on Fermi's progress.

Leo Szilard was a staunch leftist. It is never easy to know what motivates anyone, and especially Dr. Szilard. I have been thinking about Szilard since 1925, and I am afraid that I still do not fully understand him. But I think Szilard was leftist for at least two distinct reasons: First, he truly felt that communism or some similar political system would bring mankind happiness. Second, he felt that the leftist position would triumph eventually and win him a high political post. Szilard never won that high political post, nor even came close.

No one knows better than I do how many unfavorable things can be said of Szilard. But let me repeat: I liked him. He had remarkable vision, intelligence, vigor, and eloquence. He had rare gifts of understanding. Though his life was always quite busy, Szilard made clear in his own way that he sincerely cared for my happiness.

You can seek power without thinking how to use it, and that is what Szilard did. That he never held any real power was a very great disappointment to him. But what little power Szilard obtained, he did not misuse; so the usual criticisms we make of power lust hardly seem to apply.

What weakness made Szilard think so highly of himself I never knew. But for a man so conceited, it was strange how much he relied on me and how much I trusted him. I told Szilard that I wished he had less power lust. I teased him by calling him "the General." But I never pressed the point, for I knew that my words would have no effect.

A man's principle desires and inclinations are fixed properties. Even his closest friends cannot change them. Szilard lacked some essential perspective on himself. But even if I had been wise enough to fully provide it, Szilard could not have accepted it. Michael Polanyi had taught me years before that wisdom cannot be taught except to those who already tacitly possess it.

Szilard was independent, yet relied on friends. He once wrote me from England that someone should look after him because he could not do it himself. He came to stay with you because he missed you and hated cooking for himself; then he left because the bed was too hard and he was tired of you. It was sometimes hard being Szilard's friend. But it was always deeply interesting, and that depth of interest is what I try to recall.

Why does a solitary walk always focus my mind? This is a question that has intrigued and delighted me since childhood. It is quite nice to have an office and even nicer to have a warm, well-furnished home. But my mind often comes to a standstill after some hours indoors. So I take a walk. Once outside, my mind immediately begins to move freely and instinctively over my subject. Ideas come rushing to my mind, without being called. Soon enough, the best answer emerges from the jumble. I realize what I can do, what I should do, and what I must abandon.

I took many walks at the Metallurgical Laboratory. They were pleasant walks, but they were also a duty. Difficult problems always gave me walks, and we had many difficult problems.

One problem we never foresaw: that some of the chain reaction's undesirable fission product would have an absorption large enough to stop the reaction. When the first large-scale chain reaction was attempted, this soon occurred. A fission product, xenon 135, was produced, with an enormous cross section. It absorbed neutrons vigorously and the chain reaction halted.

Fortunately, the wonderful John Wheeler was working on the Manhattan Project. Wheeler swiftly analyzed the problem and saw how to overcome it. There were many similar acts of ingenuity during the war.

Fermi established the chain reaction on December 2, 1942, as I have described. Leo Szilard was able to bring himself to attend the event. His sense of history was apparently greater than his jealousy of Fermi.

Just two weeks later, the plans for the reactor plant in Hanford, Washington were passed out. They reflected the will of the Army to build production-scale nuclear reactors. These plans had clearly been drawn up well ahead of time. Already, the Army had acquired a great deal of land in Oak Ridge, Tennessee and committed millions of dollars to these and related enterprises.

What confidence all of us had in Enrico Fermi's team. He inspired it. He deserved every bit of it.

# A Squirrel in a Cage

𝓑 uilding an atomic bomb was not easy. It required the design, development, and production of a great many materials. The work was on a vast scale and had to be done quickly. In 1942, General Groves looked for a large industrial firm to perform much of this work. He found the DuPont Corporation and asked for their help. Resisting General Groves was never easy when he wanted something, and during the war it was nearly impossible. DuPont soon agreed to complete the job as best they could.

Atomic bombs were still fantasy in 1942. Large American industrial firms then did not yet understand nuclear energy. Under such primitive conditions, DuPont was as good a choice as any to perform the exacting tasks required.

Contrary to some accounts of the Manhattan Project, not all of DuPont's work was good, nor were its relations with the scientists always smooth. But in wartime, such things must be accepted.

American chemical engineers of that day got a far narrower, more "practical" training than did their European counterparts. So the DuPont staff knew less atomic theory than men from a comparable European firm would have known.

I had begun as a chemist myself, and through a lot of hard work had made myself a nuclear physicist. At first, I expected DuPont to do as I had done, to master nuclear physics. But I soon saw that I would be disappointed in this, and perhaps my expectations were unreasonable. There is something mysterious in the process of absorbing advanced physics: an emotional side next to the intellectual one. Such learning takes time. And DuPont did not have the time to learn nuclear physics properly. Time was always short during the war and all of us were constantly afraid of the weapons Germany might be building.

DuPont made drawings for us and we reviewed them. A proper technical drawing has the air of an assertion, but DuPont's drawings were more like proposals. DuPont knew that we approved each drawing before it was done. They expected us to catch the mistakes.

So the nuclear scientists in the Metallurgical Laboratory had to tell DuPont how thick the holes should be in graphite, how much water should come through, and so on. I often thought that we ought to have been placed formally on the DuPont team. DuPont seemed to like the water-cooled nuclear pile that I had sketched out with Gale Young and several others. But they never asked any of us to join the group making the detailed design.

Let me point out that DuPont did some things very well. They knew how to organize and manage a huge construction project. They knew procurement and how to obtain plenty of graphite. They knew the principles of conventional engineering. They knew how to hire people, how to organize them, and find lodgings for them.

The Manhattan Project was truly an enormous project, and such organizational skills were crucial to its success. I wish that my group had better appreciated those contributions at the time. We were almost completely preoccupied with the design aspects.

DuPont also made one design decision which later proved to be brilliant. *Accidentally* brilliant it is true, but brilliant nonetheless.

My group had designed the Hanford reactor as a large graphite cylinder containing 200 tons of uranium. DuPont squared off the cylinder so that when the reactor was built, it contained 260 tons of uranium. DuPont felt that the larger the reactor, the more room for maneuver if nuclear calculations were faulty.

And indeed these calculations *were* faulty. Our unsuspected fission product, xenon 135, absorbed so many neutrons that had the 200-ton reactor been built, the plutonium production would have been halved.

So in this case, DuPont's plan was wonderful.

By early 1943, DuPont had agreed to our water-cooled reactor design. In February 1943, with the help of Alvin Weinberg, I chose a pile-building plan that would be reliable and save precious metal. We felt the graphite block should measure 24 feet on a side and weigh about 1500 tons. We expected the pile to become critical with about 60 tons of uranium.

In February 1943, quite unhappy with DuPont, I visited Arthur Compton and offered to resign from the Metallurgical Laboratory. Compton seemed not to have expected this at all, but he handled himself well. He was unable to completely contain his emotions, but he remained, for the most part, calmly logical. He told me that I was an integral part of the Met Lab and that everyone involved in the project knew it.

Compton never tried to deny the troubles we had been

having. Just the opposite. He even made me feel—subtly, and I cannot say just how—that many of the things that disturbed me were just as disturbing to him.

But Arthur Compton took care not to castigate anyone in the process. He showed sympathy for all parties. We had a vigorous but friendly talk. He made sure that I had expressed my frustrations fully. He made no elaborate promises but he said what he could. Then he told me to take one month's vacation and return to my work. And that is just what I did.

Vacations are a wonderful thing. "Important" people ought to be persuaded to take them more often. Away from the pressure of daily work, your mind runs continuously and more freely than it is accustomed to doing. You forget some of your prejudices. You have the leisure to sift among your favorite ideas: Which of these notions are truly valid? Which are not, and why not? And your very best ideas—how might they be applied more generally? That is often the most fruitful question.

I had a lively, playful vacation, entertaining such ideas. When I returned a month later, I reminded myself of the goals of Adolf Hitler and resumed my work. Later, I heard that while I was gone General Groves had tried to ensure that I would have a greater design role when I returned. Groves was quite a boss.

✻     ✻     ✻

People have often accused me of great politeness. Well, my parents taught me to be correct and friendly, to follow others going through a doorway, and never to claim more credit than I deserve. Seventy years after leaving the home of my parents, I still try to act as they taught me. None of that can be very unusual. But I often contradict the idea that I am polite, and that already shows poor manners. I generally do not defer to my questioners; if they know something better than I do, they will not consult me, I trust.

So I do not think I can fairly be accused of undue modesty. Modesty is a complicated property, not easy to define. I would define modesty as a failure to demand important things that you clearly deserve. By this standard I am not modest and my behavior with DuPont proves it.

Discord is at times unavoidable. It would have been far nicer for the physicists in my team to chat pleasantly with the DuPont Corporation all through 1942 and 1943. But sometimes I felt bound to vigorously oppose their basic attitudes and even their specific ideas. This I was quite willing to do.

A few historians may enjoy combing through our troubles with DuPont, defining and cataloguing them, assessing blame. I do not. What is most important is that our partnership remained intact and that the nuclear reactor that DuPont built eventually made plutonium. Success always soothes hard feelings.

Sometime after the war, DuPont hired me briefly in connection with a Savannah River nuclear plant, and in the course of that work I came to understand much better the nature of DuPont's position during the war.

Indeed, I was surprised to find that I enjoyed working for DuPont. In my mind, I gladly forgave DuPont for all that had occurred during the war. But in my heart, I could not quite forget, and I find that no matter how old I get, my distaste for DuPont occasionally surfaces.

❧       ❧       ❧

Now, before I go further, let me tell you a bit of the early life of Enrico Fermi. Fermi was a year my senior, born in

Rome in 1901. His father worked for the Italian State Railway. His mother had once taught school. Enrico was the youngest of their three children. His devoted brother, Giulio, was just a year older, and together the two boys explored the mechanical world. With quite limited materials, they built electric motors that actually ran.

When Giulio Fermi died suddenly at age 14, Enrico turned to books, especially on physics and mathematics. He also became close friends with the young theoretical physicist, Enrico Persico.

After high school, Fermi won a scholarship to the Sevola Normale in Pisa. He outgrew that venerable institution in four years. His doctoral dissertation, written at age 20, was far above the heads of his teachers.

Fermi then took a fellowship to Göttingen but was unhappy there, feeling ignored by the other top students and faculty. I have no idea why this occurred. It was quite unusual for a promising young physicist to dislike Göttingen. Perhaps Fermi was somewhat proud and lonely for the company of Italians.

After Göttingen, he went on to the University of Leiden, where he was greatly encouraged by Paul Ehrenfest. Fermi later taught in Rome and Florence.

Then Orso Corbino, the chairman of the physics laboratory at the University of Rome, arranged for Fermi to come to that university as a full professor in theoretical physics. He was then only 25. In 1929, he married the former Laura Capon, a university student.

Fermi's team of physicists bombarded uranium with neutrons in 1934. It caused fission, though neither he nor anyone else realized it for another five years.

Enrico received the 1938 Nobel prize in physics without wearing a fascist uniform or giving the fascist salute. The fascist government demanded universal support, and when Fermi failed to accept the Nobel prize as a fascist, many Italians were annoyed. Enrico was annoyed by the fuss they made; it was just the sort of thing he disliked.

Fermi and I had been given real American names for the war. Mine was "Eugene Wagner" and Enrico's was "Henry Farmer." One day we were driving together along a high-security road. At the checkpoint, the military guard asked my name. I said, "Wigner—oh, excuse me please!—Wagner." The guard could not help but notice my Hungarian accent. He regarded me with suspicion and asked sternly: "Is your name really Wagner?"

What could I say? I had no idea. But Enrico saved me. Quite firmly and confidently, he said, "If his name's not Wagner, my name's not Farmer." And the guard let us pass. That quick self-assurance was so typical of Fermi.

When people discuss the Manhattan Project, they often linger too long on the physicists. We had some wonderful chemists too, whose work was essential to our success. Glenn Seaborg was my favorite: a tall, friendly, brown-haired real American.

The nuclear chain reaction made plutonium, which had to be separated from the uranium in the reactor. With talented men assisting him, Seaborg did much of this separation work. After the war, Seaborg shared the 1951 chemistry Nobel with Edwin McMillan for related work with transuranic elements.

James Franck, Edward Teller's mentor in Göttingen ten years before, worked on radiation chemistry with his usual grace. George Boyd worked on analytical chemistry; a man named Coryell studied the chemistry of fission products. Frank Spedding was another able chemist on the project.

I was sorry to have so little contact with Seaborg, Franck, and the others. I knew enough chemistry to have helped them. And they would have taught me a great deal. But my group was busy with the reactor. So we left the chemistry alone.

My old friend from the University of Wisconsin, Gregory Breit, worked on the Manhattan Project, too. In the first months of 1942, Breit coordinated the different fast neutron research projects at various American universities. Breit also understood some of the special problems in making the fissionable uranium 235 or plutonium 239 into a bomb. In early 1942, Breit and I together tried to decide: What size sphere of uranium would best support a chain reaction?

Around May 1942, Gregory Breit resigned from the Manhattan Project. I never knew why; he had probably argued with some people, become enraged, and then chosen to walk away from the source of the conflict. There were many things about Gregory Breit that one never really learned. Breit believed himself a natural leader. Perhaps he was. But when provoked, he had a violent temper.

When I spoke of Breit, other physicists would say, "Oh, the quarreler. What of him?" Indeed, Breit *was* a quarreler, but it is unfair to dismiss him with one phrase. How many quarrelers are unselfish? I was lucky to have seen closely Breit's unselfishness at the University of Wisconsin. And I was sorry to see Breit leave the project. He remained as a consultant through 1944.

❈          ❈          ❈

I am a poor judge of leadership talent. Part of the reason I say that even Gregory Breit and Leo Szilard might have made good rulers is that at first I underestimated a number of the top leaders in the Manhattan Project, including General Groves and Robert Oppenheimer. But the man I most underestimated as a wartime leader was Arthur Holly Compton.

We all knew of Compton's intelligence and achievements in physics. But at first I thought he was a poor choice to run the Met Lab. You see, Compton had never been a nuclear physicist, and physics was no longer even his main interest. So he could hardly give technical advice. Worse than that, he seemed at first withdrawn, when we desperately needed a vigorous promoter of our atomic bomb program.

But Arthur Compton became a superb leader: friendly, sincere, and supportive. He knew how to recruit talent; he trusted immigrant Americans as able, politically reliable men at a time when many Americans did not. He understood human desires: how to accommodate them, how to blend them into a working team.

Compton spoke clearly about what he wanted, and he was careful to provide new challenges as soon as the old ones were solved. He tracked our problems carefully and did his best to maintain good relations with DuPont. Compton knew that freedom, independence, and imagination were essential in the Met Lab. And he smoothly handled his superiors so that the scientists were never ordered around like military privates. That is a superb set of qualities to be found in one leader.

Arthur Compton had very few flaws. Some people felt he was overly religious and described him with phrases like "missionary zeal." His critics seemed to resent certain religious missionaries who helped mankind but insisted on imposing a religious framework on all progress. They said that Arthur Compton was a bit that way.

Well, I have been impressed by the religious missionaries I have met, but I must say that Arthur Compton did not much resemble them. I found Compton neither overly religious nor overly zealous in any way. In fact, I wished at times that he had been more zealous about mastering the mechanics of particular pieces of machinery. He found his duties too broad to allow it.

Arthur Compton's only major flaw was a stout conservatism. He felt it his duty to discourage what he considered crazy ideas. Making nuclear energy from the combined explosions of small bombs, for example, was not the conventional project of an older physicist. So Arthur Compton did not favor it. He never tried to kill such ideas, but he diverted resources from them, which had a similar effect.

❧        ❧        ❧

Not all of the great physicists in Germany had left when the Nazis took power. Otto Hahn, Max Planck, Werner Heisenberg, and Max von Laue remained.

Otto Hahn had joined the Kaiser Wilhelm Institute back in 1912. In 1928, he had become its director and stayed there through the war. Dr. Hahn was one of the world's top experts on fission; after the Second World War, he received the Nobel prize in chemistry for his discovery of the fission of heavy nuclei.

Werner Heisenberg remained in Leipzig until 1941, teaching theoretical physics, as he had taught Edward Teller. Then he came to Berlin as a professor at the University of Berlin and director of the Kaiser Wilhelm Institute for Physics. He stayed there nearly through the war, working with Hahn to develop a nuclear reactor. Planck and von Laue also held important scientific posts through the war.

None of those men were instinctive Nazis. They disliked the Nazi ideology; they did not want Hitler to rule the earth.

Just the same, they were Germans and they were patriots. They felt quite rightly that it would turn their country upside down if the Germans lost the war. They did not want that to happen. So they attended to their own nuclear bomb project.

One day while we were working at the Met Lab, a telegram reached us from a fine theoretical physicist named Fritz Houtermans. Houtermans understood fission; he also knew Heisenberg quite well. Houtermans wrote from Switzerland: "Hurry up. We are on the track."

I believe that Werner Heisenberg purposely slowed his wartime work, hoping that the atomic bomb would not be realized. I have no evidence of this beyond what I know of Heisenberg's character. But character is the crucial element in a man's life. I cannot imagine Werner Heisenberg gladly handing Adolf Hitler an atomic bomb.

It turned out that our fears of a German atomic bomb were unfounded. Whatever their motives, the top German scientists were missing crucial materials for building a bomb. They never approached a working atomic bomb. And Hitler never pressed them to do so, believing that an atomic bomb would take many years to build, by which time he would already rule the earth. Well, Hitler was right to think that the bomb's development would take years. But he did not see how long his World War would last.

Now I am an old man, and I am startled to realize that Hitler has been dead for over 45 years. Now I can say: "Dictators are rarely technical innovators. We should have known that Hitler would not build an atomic bomb."

But in 1941, this was not at all clear. Hitler was then the master of the novel blitzkrieg strategy. He seemed a quite modern man. I think I was quite right to worry in 1941 that Adolf Hitler was building an atomic bomb.

By early 1944, the pressure of the work had begun to

subside. The war was going better for our side. Hitler showed no sign of having an atomic bomb. We recovered our balance and began to survey the likely effect of our atomic bomb and of a world with atomic bombs.

You cannot hide huge atomic installations for long. We hoped and sincerely believed that the President and the U.S. Congress would have to adjust all of our war policies to the new reality of the bomb.

James Franck, in writing his Franck report, suggested an international agreement to ban nuclear arms and asked President Truman to demonstrate the bomb before using it. Robert Oppenheimer spoke of sharing the secrets of the bomb with the whole world.

We had built the atomic bomb in fear that Hitler had one in Germany, and perhaps his Italian friends had one, too. That was the pressure. We needed an atomic bomb ready to drop on Germany, or at least to scare them with if they seemed ready to drop one on an Allied country.

We had little doubt that the atomic bomb would someday be built. Far better that it be built by the United States than in Hitler's Germany; better for the United States and better for the world.

By late 1944, we knew that the Germans were not building an atomic bomb. Some people ask, "Why didn't you halt work on our own atomic bomb as soon as you knew this?" But by then the work had a great momentum, an insistent military boss in General Groves, and a dynamic scientific leader in Robert Oppenheimer.

Oppenheimer, the head scientist at Los Alamos, is another man whose deeds should be far better known. His name is fairly well known today, but most of the popular ideas about him are false. Many histories and biographies have called Oppenheimer "arrogant," just as they delight in calling Edward

Teller arrogant. Every man has a touch of arrogance in him, including these biographers. Oppenheimer was a curious man, and people not easily sorted and categorized are often labeled "arrogant."

I never saw real arrogance in Robert Oppenheimer. He was attractive and expressive. He knew a good bit about himself and was quite interested in his appearance and his reputation. Is that arrogance? If so, then the world is largely populated by arrogant people.

Oppenheimer's only temperamental failing was a tendency to withdraw from social company. At large meetings, he sometimes stood alone in a corner, solving some private problem. But mostly, he was lively, friendly, well relaxed, and intent on the crucial issues. He was a very busy man who nearly always found time to consider the well-being of others. And that is the very opposite of arrogance.

Oppenheimer dressed formally; I cannot ever recall seeing him without a suit and tie. Many people have described his "piercing gaze." He had fine blue eyes, yes, but I never felt them pierce me. He was a little above normal height and unusually thin. Some people said that he had shriveled through overwork, but this was his natural build.

You know, most bosses have quite ample bodies. A large body boosts a man's physical presence. General Groves's body certainly made a strong first impression. And I have found that most directors do not lose weight under stress, they gain it. They stay up nights snacking, attend richly catered receptions, and so on. So Robert Oppenheimer must have been naturally thin. If his build was a handicap, he accepted it without comment.

I had known Oppenheimer slightly for many years before the war. Nearly every physicist felt that he knew Oppenheimer at least a bit; he was the kind of man who is talked about.

Oppenheimer did things very well, knew a great many physicists, and left a vivid impression. I always thought of Oppenheimer as a European; his gestures and temperament were much closer to the colloquia and coffeehouses of Berlin than to anything American.

I would have liked to have seen Oppenheimer far more often in the late 1930s. But he was in Berkeley, all the way out in California. The distances in the United States awed me; once I went out from New Jersey to Chicago, got lost several times on the way, and vowed I would never take such a long trip in America again. Physicists did not ride airplanes in those days, so before the war Robert Oppenheimer was always four days by train and a great many dollars away.

But Teller, Szilard, and I would see Oppenheimer here and there at a physics conference. We all thought he was a bit crazy. Those were the years when we most feared Hitler would conquer the earth, and Oppenheimer seemed to feel that Hitler was not even dangerous! Whenever Teller and I would mention Hitler's latest actions, Oppenheimer brushed them off. He wanted to discuss physics.

Oppenheimer had leftist ideas and favored a stronger, more centralized government. He did not want to see the United States go communist, but he sympathized with communist regimes in other countries.

I do not attack him for those beliefs. Unlikely as it seems today, communism was quite popular in the 1930s. I opposed it because I had seen it seize power and property. But the deepest evils of communism were unclear in the 1930s, even to a conservative like Wigner. Oppenheimer was slow to grasp simple political truths. But once he understood them, he was completely reliable. Robert Oppenheimer was never a security risk.

Los Alamos had perhaps the hardest role to play in the

Manhattan Project. They took most of their materials from various other weapons plants. But they had to decide how to explode this material, how to build a bomb that would fit in an airplane, how to fuse it so that it would detonate properly, and many other difficult questions.

Los Alamos did all this superbly, and Robert Oppenheimer deserves much of the credit. He was recruited by General Groves to run Los Alamos in October 1942, and from then until the end of the war he kept Los Alamos running as smoothly as any leader could have.

It is easy to criticize a man with Oppenheimer's sensitive temperament and unusual habits. He did not look like a proper leader. He held himself slightly apart from others. But he *was* a good leader. He led the work at Los Alamos with grace and flair, and made very few mistakes. Nearly everyone who worked there during the war agrees on that.

Oppenheimer was famous for his facility with foreign languages and his sweeping cultural interests. But I never knew him intimately enough to share them. Except for some strained political talks in the late 1930s, we barely discussed topics outside physics.

But certain things about Oppenheimer were clear even to a casual observer: that he was an able man who liked people, a strong boss who was wise enough to see the world as it is, and flexible enough to adjust to it. The men at Los Alamos were highly intelligent and independent people. They disliked being visibly directed. Oppenheimer understood that. He knew their strengths and weaknesses without asking and treated them with some sensitivity.

Oppenheimer could be quite deft. He smoked a pipe and he gave some deft direction with his pipe alone. When a subordinate reached him with a grievance or request, Oppenheimer received him with the pipe in his mouth. He listened carefully to the man, all the while making clear that his pipe also required some attention.

In this way, Oppenheimer quietly sent his subordinates an important message: that a successful project understands the personal needs of its members but does not cater to them. He did all this very easily and naturally, with just his eyes, his two hands, and a half-lighted pipe.

Most workers want to be catered to and in their hearts feel that they deserve it. That is human nature. But it is the boss's job to say no, to value the product of the company over the happiness of its workers.

And so it is the fate of most bosses to grow unpopular. I saw it happen to my father in the leather tannery in Budapest. But it did not happen to Robert Oppenheimer; he was most popular with those who had worked with him for many years. And that is quite an achievement for a boss.

Yet people still ask me, "Oppenheimer—quite an arrogant fellow, wasn't he?" Yes, Oppenheimer had some arrogance. Let us be glad he did; he needed every bit of it for his work. There is a German proverb: "Modesty is an ornament. But one gets farther without it." And I think Robert Oppenheimer knew the truth in that proverb.

"But was he always consistent?" people ask me. No, perhaps not. But as Bismarck said, only oxen are consistent. No one ever mistook Robert Oppenheimer for an ox.

❊    ❊    ❊

During the war the Army did not want us to know everything about the hydrogen bomb or even the atomic bomb. We were told to consider only the specific role we had been chosen for—in my case, to produce plutonium and nuclear reactions.

Physicists like me were not supposed to know that the

Los Alamos work even existed. But we certainly knew about its work in areas like our own. Someone once addressed a letter to me this way: Eugene Wigner, Nuclear Energy Project, University of Chicago. The letter arrived promptly.

Physicists need to discuss their work, and not only because gossip is very human. Scientists do not like to deceive their friends and colleagues. And they know that discussion improves their work.

So we in Chicago knew about Los Alamos and about the chance for a hydrogen bomb. But we also knew that the political and military bosses wanted us to ignore the hydrogen bomb and everything beyond our own work. Most of us did. And in 1943 the emphasis of the Manhattan Project shifted toward Los Alamos.

Edward Teller worked on the hydrogen bomb during the war. He was already intrigued by it, and I discussed it with him then. I did not want to work on such a bomb, and Teller hardly tried to convince me to do so. But he knew that I approved of his own work.

When it seemed clear that the Met Lab would create its large-scale nuclear chain reaction, Teller turned his mind to how to use an atomic bomb. He moved to Los Alamos to work under Robert Oppenheimer. In August 1944, Enrico Fermi also moved to Los Alamos.

Teller's years at Los Alamos were not happy ones. Secluded from most of the people and things that give life meaning, their work rigidly organized and discussion nearly forbidden even with close friends—this hampered all the physicists there. Secrecy is deeply averse to human nature, especially when imposed by others. Suppressing real intellectual excitement is quite painful. And Teller, who has an almost physical need to discuss his work, was specially pained. It is a tribute to Edward that he worked so brilliantly all through his unhappiness. His work with shock waves and matter under high pressure still shapes those fields today.

Germany finally surrendered to the Allies in the spring of 1945. Adolf Hitler killed himself that April in his bunker in Berlin. I could not help but feel that the act of suicide, in most lives so grotesque, was appropriate in Hitler's case. I found that I could not forgive him for the millions of people he had killed, and over time I have stopped trying.

The United States dropped atomic bombs on Hiroshima and then on Nagasaki. On August 14, 1945, the Japanese also surrendered. The Allied forces had finally won the Second World War. Germany had lost over 4,000,000 people, counting the Jews killed by Nazis. Poland had lost about 6,000,000 people, more than half of them Jews. Soviet Russia had lost about 11,000,000 soldiers and 7,000,000 civilians. By comparison, the United States had lost very few: about 300,000 soldiers and a few thousand civilians.

Very soon after the war, Robert Oppenheimer felt guilty about his work, about the legacy to the world of the bombs dropped on Hiroshima and Nagasaki. He often spoke about his guilt. Many other scientists felt equally guilty, but Oppenheimer's guilt is far more famous because he expressed it more profoundly.

The end of the war called for new thinking—diplomatic, political, and military. Like most physicists, I thought: "Why should we refine this terrible bomb now? The war is over. Basic physics is far more pleasant and intriguing than hydrogen bombs." So many physicists reasoned this way that Los Alamos was nearly abandoned.

Szilard had begun to forget about bombs as soon as the atomic bomb was near completion. Once the success of a venture was assured, Szilard lost interest. That was the way he was. So Szilard went off to work on the breeder reactor.

Teller left Los Alamos as soon as it was graceful to do so. But he returned to military science soon after. He had opposed the policy of bombing Hiroshima without warning, but he was deeply disturbed by the great exodus from Los Alamos. Teller felt that the defense of his adopted country was critical, that the hydrogen bomb could be perfected, and that he should finish what he had begun. So he devoted himself to developing the hydrogen bomb.

❋     ❋     ❋

Ever since the Second World War ended, people have asked me: Knowing what I do now, would I again be willing to help build the world's first atomic bomb? I would like to say that I regret working on the bomb, if only to please most of my questioners. But I cannot honestly say that I do, either intellectually or emotionally.

In fact, I wish the bomb had been built sooner. If we had begun serious fission work in 1939, we might have had an atomic bomb ready by late 1943, when Stalin's army was still bottled up in Russia. By August 1945, when we first used the bomb, Russia had begun to overrun much of Central Europe. If we had held an atomic bomb in 1944, The Yalta Conference would have produced a document much less favorable to Russia, and even Communist China might have been set back. So I do not regret helping to make the atomic bomb.

On the other hand, I never wanted the bomb dropped on Japan. What a wrenching decision! Once Germany surrendered, I hoped the atomic bomb would be made public. I felt hiding it would only hinder our political transition to the day when such bombs were common.

Like most of my colleagues, I did not expect the bomb to be dropped on Hiroshima and Nagasaki. With Hitler decisively defeated, I wanted to see our leaders consult an international panel before using the bomb. And I would have opposed using the bomb on Japan unless that panel had given their consent.

But the Manhattan Project had been a project of the U.S. Army. So perhaps we should have known that the bomb would be used as a drastic military weapon, not a foreign policy instrument. The Army wanted quick results. They were not sure that these bombs would explode and did not want to be embarrassed if they failed. The Army also seemed to feel that a prior demonstration of the bomb would greatly reduce its psychological effect.

So on August 6, 1945, the U.S. Army dropped an atomic bomb on Hiroshima. Three days later they dropped another on Nagasaki. I was very sorry to see our country set a precedent for using atomic bombs as regular weapons of war.

Now, modern warfare may be no worse than ancient; Germany lost more than a third of her people in the Thirty Years War between 1618 and 1648. But the atomic bomb was a horrible new method of killing.

Though the Hiroshima and Nagasaki bombs forced Japan to surrender and actually killed fewer people than would likely have died in an American invasion of Japan, I still cannot feel that dropping these bombs was a good thing. They killed not only soldiers but a great many innocent civilians.

One of the most curious things about the atom bomb was that its very evil made many of its creators feel a certain elation. We assumed that a weapon this evil would force all of the world's most powerful nations to unite, at least in part; to submit their military forces to an international authority like the United Nations. We seriously hoped that in time all nuclear weapons would be abolished, and perhaps even all warfare.

Before long, most of us saw the vanity of such hopes. We could neither neutralize the force of such weapons nor restrain the appetites of nations pursuing them. And we could not blame any country for trying to build a bomb we had developed ourselves. So the atomic bomb became a tremendous disappointment to me.

Today it is puzzling that so many able people could have seriously hoped that the atomic bomb would bring world peace. But it is important to recall that we did. In fact, the hope of ending all wars was almost as great a spur to our endeavor as was our fear of Adolf Hitler.

Now the war was over, but it was still hard to look ahead with confidence. Responsible people still spoke of keeping the bomb secret and creating an absolute defense against it. We had already made too many mistakes in the 1930s. It was too late to make any more.

Apt lines were written by the Hungarian poet Vörösmarty:

> Nor lives the world for ever
> But while it lives and turns, it is not idle.

# Fifteen

# *"Isn't He the One Whose H-Bomb May Blow Up the World?"*

*T*he war changed everything for physicists. After the Second World War, physics grew so fast that no one could know the whole picture anymore, or even anything close to it.

After Hiroshima, American physicists spoke with new confidence and a new authority. University faculties grew, so we knew our colleagues less well. The federal government began funding a lot of research, especially in high-energy physics. Salaries were higher and research more tightly organized.

All of this was related to the fact that scientists were no longer considered crazy. Just the reverse: We were now treated as experts on many diverse issues that most of us had scarcely thought of before the war, such as the nature of modern warfare and prospects for disarmament or world government.

It was much harder now for scientists to be shy and retiring. For the first time, it became reasonable for men seeking power and political influence to plan a career in science. The veterans of the Manhattan Project now had some power, perhaps even more power than they had knowledge.

It may have been healthy for scientists to hold political and social power, but it was certainly different. Before this, most American scientists had ignored public life. Benjamin Franklin had been a great public servant and a great scientist. But Franklin kept science in a separate compartment. He was not a public servant as a scientist.

By 1946, scientists routinely acted as public servants *as scientists,* publicly addressing social and human problems from a scientific viewpoint. Most of us enjoyed that; vanity is a very human property. We joined panels, held conferences, and speculated grandly about the future.

I had mixed feelings about all this. Certainly, scientists had the same political rights as other citizens. We had the right and perhaps even the duty to speak out on vital political issues. But on most political questions, physicists had little more information than the man on the street. Yet the pace of public debate, our excitement with our new status, the popular fear and awe of the atomic bomb—all these things led physicists to speak like political experts.

In 1946, I helped organize a conference at Princeton called "The Future of Nuclear Science." We brought in physicists from all over the world. We worked hard to get some even from Russia. Some fine theoretical papers were given, but the conference was more political than scientific. The two great, obvious questions were political ones: "What are we going to do with this atomic bomb?" and, "Might atomic bombs be used to abolish warfare?"

Those questions are still unanswered today. A whole generation of political leaders has evaded them. Perhaps the next generation will do better.

❊    ❊    ❊

When I first entered physics in 1921, people used to smile when I said I was a physicist. They saw my profession as the harmless pursuit of complex irrelevancies. Now they had stopped smiling. I had some pride, and I liked that.

But I was one of many scientists who also looked back fondly to the days when science had been a monastic calling. One scientist wrote a song that expressed our feeling well: "Take Back Your Billion Dollars." All of this money had brought bureaucracy and taken some of the pleasure from the practice of science.

Modern physics was also disturbingly specialized. Specialization is productive; I clean the house much less well than my wife, so she cleans the house while I practice science. Scientists who specialize can pay closer attention to their work and better master it.

But it is sad to lose touch with whole branches of physics, to see scientists cut off from each other. Dispersion theorists do not know axiomatic field theory; cosmologists do not know nuclear physics. Quantum mechanics is hard to explain to a chemist; its terms and concepts are highly developed. And yet the best theoretical chemists really ought to know quantum mechanics.

Specialization of science also robbed us of much of our passion. We wanted to grasp science whole, but by then the whole was something far too vast and complex to master. Only rarely could we ask the deep questions that had first drawn us to science.

In 1870, the year my father was born, a first-rate physicist could expect to master every branch and aspect of physics. There was a great ignorance surrounding many basic physical elements, but there was also a freedom and graciousness that allowed physicists to range freely over the field.

Even as a young man in the 1920s, I had expected in my heart to one day know all of physics. I was ashamed then that I hardly knew planetary theory or electromagnetic theory. I said, "Well, I will begin to remedy the situation just as soon as I finish writing this article . . ." Perhaps the expectation of learning all of physics was just an illusion, but it was a beautiful illusion, and near enough to the truth to be credible.

One day around 1942 I told James Franck, "I don't think I will give much to physics after the war." Physics is a young man's game; I was then 40 years old and beginning to feel like an old fogey. Franck disagreed with me, but only because he felt that physics would evolve slowly after the war. But the growth of physics never slowed—it sped up. And that changed physics, not only in practice, but emotionally as well.

By 1950, even a conscientious physicist had trouble following more than one sixth of all the work in the field. Physics had become a discipline, practiced within narrow constraints. I tried to study other disciplines. I read *Reviews of Modern Physics* to keep abreast of fields like radioastronomy, earth magnetism theory, and magnetohydrodynamics. I even tried to write articles about them for general readers. But a growing number of the published papers in physics I could not follow, and I realized that fact with some bitterness.

❈　　　❈　　　❈

What I saw in Europe after the war was even more bitterly discouraging. Russian leaders dreamed openly of subduing all the earth. The Second World War had extended Russian territory by some 250,000 square miles and 23 million people. America took neither territory nor peoples in the war. We liberated territories, then withdrew.

Between 1920 and 1940, I had largely immersed myself in physics. Between 1940 and 1945, I had learned something about the technology and administration of modern warfare. Now, for the first time, I could study politics closely.

I was dismayed. There seemed to be slogans and brutality on the Russian side; slogans and nonsense for the Americans. Slogans please the heart, but they are a way to avoid thinking. And I wanted the American people to think seriously about Soviet aggression in Europe and how best to stop it.

We had a monopoly on nuclear weapons for a few years and very small risk of effective retaliation. Though we were often provoked, we did not press our military advantage. The Russians blockaded Berlin, occupied Czechoslovakia, and broke the Hungarian Peace Treaty. Russia took Estonia, Latvia, and Lithuania, and the United States did next to nothing about it.

Perhaps the United States could not have risked sending soldiers to Eastern Europe. But I felt we could have done far more to help the democratic elements there to resist Communism. They needed weapons, money, and publicity; the United States was a large producer of all three of those things, but we gave them sparingly to Eastern Europe.

In 1990, the Russians at last officially admitted their mass slaughter of the Polish at Katyn in 1939–1940. After 50 years, they admitted killing about 15,000 captured Polish soldiers. Very good. But in 1950, Russia fiercely denied all this. She kept her people as slaves, sealed her borders with machine guns and barbed wire, bullied her neighbors, and had the brashness to scold American democracy!

This went on for years. Mr. Khrushchev forecast that

capitalism was "doomed to destruction." In November 1956, he told Western diplomats, "Whether or not you like it, history is on our side. We will bury you." Later Khrushchev met with our president and said, "Don't worry. If I offer my embrace you will not refuse it." So he knew very well that the United States was not essentially an aggressor. Why then did he keep building his armies?

Just as no Frenchman could admit in 1938 that Hitler might overrun France, no American could admit in 1950 that Stalin might overrun the United States. Yet Stalin's dream of world conquest was just as public a dream as Adolf Hitler's. At least in 1938, France had been supported by strong allies. What country could help the United States, I wondered, if it were bombed or invaded?

The Russians vigorously practiced science after the war, but science only in the service of their military. I recall meeting with a Russian scientific negotiator in 1958, to verify the competing claims of our two governments on a technical issue with clear military implications.

This kind of meeting is very hard for an American scientist. You begin as a healthy skeptic of both governments while your Russian counterpart holds a healthy skepticism of the American government only. He has been assigned by the Communist party to military research, and his career depends on properly advancing the official Soviet view. This creates in him a single-minded intensity that is hard to resist.

This negotiator strongly hinted to me that there were things the Russian leaders would have done in Western Europe but for the United States Army. Without our military, the borders of Stalinist Russia might have moved much farther west. But even the U.S. Army did almost nothing to stop the Russians from overrunning Hungary.

The Yalta agreements had promised Hungary free elections, free government. Very good. Free national elections were held in 1945. The peasant party was called the Small Holder party. It got about 60% of the vote and most of the government seats; the Communists got about 5% and the Social Democrats 14%.

A republican constitution was adopted in 1946. The new Hungarian republic began making political, social, and economic reforms, most of them quite popular. The Russians ordered the Hungarian Communists and Social Democrats to unite. Their coalition won them a few seats more.

One day in early 1948, the Hungarian prime minister left the country to negotiate some loans in Switzerland. During his absence, the secretary of the Small Holder party, Imre Kovacs, was arrested by the Communists. After a week, they produced a "confession" that Kovacs had wanted to overthrow the government. After another week, they produced his body.

The Communists warned the Hungarian prime minister, who was still in Switzerland, that his wife and children could safely join him only if he promised not to return to Hungary. He accepted that offer. That is how communism was reinstalled in Hungary.

Industry was nationalized. The land was made more collective. By 1949, everything Hungarian came under Russian control. I did not return to Hungary for nearly 30 years. I was afraid of what I would see. You will recall that Hitler was not a man I liked. But even Hitler permitted some emigration. Russia did not permit free emigration. How bitterly I resented that!

Now, what did my newly prestigious American scientific colleagues think of all this? Sad to say, many of them seemed quite as blind to it as were the scientists of 1938 who had blithely ignored the Nazis.

❦ ❦ ❦

Physics does not try to explain all of nature, but only its *regularities.* Just as the laws in our legal code regulate actions only under certain conditions, so the laws of physics describe the behavior of physical objects but only under certain well-defined conditions. The decision to restrict the parameters of physics has been crucial to its success. I do not quarrel with it.

But it is clear that physicists work in a world of contrived rationality. And perhaps this leads them to expect rational solutions to questions that are political and military at root. If the rulers of Russia want to annex Hungary and the United States opposes the annexation, a rational solution can no more be found than when two men want to marry the same girl. One man must yield.

I loved to follow the workings of American democracy. I could not imagine a better political system. Yet I was puzzled to observe that many American citizens who refused to believe their own elected officials readily swallowed all kinds of wild lies made by foreign dictators.

My fellow scientists generally ignored Russian brutality. They never saw how vast a threat is posed by every great dictator: a threat to the freedom of whole peoples and nations and human ideas; a threat even to the idea of science.

For no dictator can be devoted to pure science. He may back certain scientific programs that hatch military weapons. But once his scientists have built those weapons, they begin to annoy the dictator. They constantly raise new questions. They want to pursue projects with no clear military value. They refuse to quote Lenin and point out that much of Karl Marx's writing is conceited nonsense.

The dictator resents all this. He is not accustomed to freethinking. He senses that free inquiry implicitly violates the absolute nature of his rule. He wonders, "How can I put these intellectual upstarts in their place?" He knows that he cannot restrict their love of science. But their freedom to practice science—that he can very readily restrict. And very often he does.

In October 1961, I attended the annual meeting of the Union of German Physical Societies. It was delightful. More than 2200 scientists met in the imperial pomp of Vienna, and many of the papers given were superb. But even then, more than 15 years after the war, not all of the scientific damage done by the Nazis had been repaired.

I have always been treated well in the United States. Yet the strength of my anti-communism has rarely been well received here. Popularity is a curious thing. People will say, "Dr. Wigner, please address our winter conference." "Dr. Wigner, please accept this honorary degree." For years, I have had more offers of speaking engagements than I could accept.

But my hosts often make clear that they value certain of my thoughts far more than others. They want to hear me discuss "Paradox in Quantum Theory." They do not want to hear me discuss Soviet brutality. Paradox in quantum theory has not killed anyone. Soviet brutality has.

People criticize me today for attending scientific conferences sponsored by the Unification Church of the Reverend Sun Myung Moon. No church doctrine shapes the scientific focus of these meetings. Leftists tell me that I should avoid them anyway because this church is fiercely anti-communist. Well, so am I. And if these scientific meetings are "tinged with politics," then I must say that every scientific meeting I have attended since 1945 has been tinged with politics. The atomic bomb made us all little politicians.

❄     ❄     ❄

My great friend Edward Teller was as upset as I was by what he saw in Europe after 1945: the deceit and treachery, the humiliation and torment of conquered peoples. Most men have a few passionate admirers. Great men also have passionate opponents. And Edward Teller has been a great man. I suppose that men like Teller are fated to live their lives on an open stage, playing to a fickle public. So I should expect to find him harshly criticized as well as admired. But after the war, when a few people began accusing Teller of greediness, ill will, and a rash temper, the criticism pained me.

Teller was not a conventional right-winger at all. He hoped deeply that someday all nations might create a world government. But until then, he felt the safety of the West lay in the forceful, unhindered development of nuclear weapons. So, after finishing his work on the atomic bomb, Teller worked even harder on the hydrogen bomb. He played a decisive role in the conception of our first hydrogen bombs in the early 1950s.

But he found that 1952 was not 1945. When we made the first atomic bomb, most people said, "Very good. Now we have won the world war!" But when they learned about the hydrogen bomb, they said, "How frightening! You may well blow up the world!" Teller was not praised for helping to invent the hydrogen bomb.

By 1952, I knew very well that political feelings are not logical or reasonable. Still, I was deeply surprised to hear people blame Edward Teller for the existence of the hydrogen bomb. Without Teller, they said, this bomb would never have been invented.

What a foolish charge! Great weapons will always be developed soon after it is clear that they can be. And the hydrogen bomb was clearly possible in 1946; it would certainly have been invented by 1960, even if Edward Teller had never been born.

Still, when most people spoke of Edward Teller, they said, "Oh, yes, isn't he the one whose hydrogen bomb may blow up the world?" The acrimony generated by Teller's political and military beliefs obscured his value as a physicist. Very few physicists in this century have made as many subtle, penetrating observations as Edward Teller.

But penetrating statements stir controversy, and Teller's conscience brought him to contrary positions. He felt that the military work at Los Alamos had become dangerously isolated from American society and that not all of American nuclear weapons research should come from one laboratory. So he pressed for a new lab. The birth of the Lawrence Livermore Laboratory owes a very great deal to Edward Teller.

Teller and I felt that military power, like police power, works best as a constant presence, persuading strong nations to respect those that are weaker. Teller widely denounced dictatorships. Nothing he said was false, but his passion was discomfiting. Anyone who sees Teller up close cannot help but observe how great is the distance between himself and the Edward Tellers of this world.

Teller had various "crazy" ideas about nuclear weapons and nuclear power. He rejected the idea of a test ban on nuclear weapons. He thought that nuclear explosives might be used productively to reshape our harbors and power household appliances. And he insisted that the health hazards of nuclear weapons fallout were a small and necessary price to pay for military readiness.

These convictions were personal to Teller, but soon observers made from them an official "school" of thought. Just as they had to anoint Teller "father of the hydrogen bomb," so they gathered his beliefs and made from them a "school."

And from the beginning, the Teller School was one with many stout critics. They said that Teller was misbehaving, that he was warlike and wished to see nuclear war. Favoring nuclear war—that was the popular definition of "the Teller School." What a crazy thing to say about Edward Teller, who has always shown such a marked aversion to every kind of violence!

People seem to need to attach important ideas to a single person, often with little foundation. Enrico Fermi is perhaps best known for the "Fermi statistics." Now, Fermi was a brilliant scientist. No one could think more highly of him than I do. But his Fermi statistics did not surprise anyone in Berlin. Many of us were already using them. I wish Fermi was well-known for his lovely theory of beta decay. But no, people remember "Fermi statistics."

Likewise, many physicists before me knew that when graphite is bombarded by neutrons, it is damaged in a certain way. But when I pointed this out, people said, "Oh, yes, very good! Let us call it 'the Wigner Effect.'" And many people still use that phrase. I no longer protest; it is a far better label than "the Teller School." But this is credit I do not deserve.

Teller, by contrast, gets blame he does not deserve. People have for many years accused him behind his back of a deep power lust. That is an easy charge to make against a forceful man. Sometimes the charge is even justified, as was partly true of my dear friend Leo Szilard. But in Teller's case, the charge of power lust is false. He has held some prominent posts: professor of physics at the University of Chicago from 1946–1952, and later associate director of the Lawrence Livermore Radiation Lab that he had helped create.

But is that too much power for one of the world's greatest physicists? Should Teller have taken a menial position somewhere where his political views would be ignored? Not at all! It hurts me to see Teller's motives so badly misjudged. I can only guess how badly it has hurt Teller himself. He is quite a sensitive man.

No one who really knows Edward Teller could ever say that his chief motive is selfishness. His chief motive is the defense of freedom. He loves truth and equality. Teller respects the logic of organizations and has taken care to express his view on science and politics through proper channels. Even when he dislikes official rules, he still obeys them. I can spot selfishness in a man, and Edward Teller is not selfish.

Photography is a worthy and intriguing medium, and a skilled photographer can create revealing portraits. But photography at this level is rare, and rarer still when the subject is a restless man, who rarely settles long in one spot. I have almost never seen a photo that does full justice to Edward Teller. His photographs clearly show an intelligent and cultured man. But they barely suggest his great concern for others and his capacity for glee. Photographs have failed Edward Teller.

And Teller has hurt his reputation by often disputing his critics. He has always relished arguments and argued with great passion. It is a bit disturbing to argue with a man of such brilliant and vehement sincerity.

Teller has always used friendly arguments to sharpen his ideas. But he has always been sensitive to the criticism of friends. Enrico Fermi once burst out in mock exasperation: "Teller, you are the only monomaniac I know who has thousands of manias!"

And apparently Teller has a temper. But in the 65 years I have known him, he has never once shouted at me. So how numerous can his tantrums be?

# *"Thank You Very Much! But Why Are You Congratulating Me?"*

When the war ended, it was time to turn to peace. In 1946, I took a job as director of research and development at the Clinton Laboratory in Oak Ridge, Tennessee. The Clinton Laboratory had been a crucial part of the Manhattan Project. Today it is called Oak Ridge National Laboratory.

Gale Young and Alvin Weinberg, my top assistants at the Metallurgical Laboratory in Chicago, were heading to Clinton, and I agreed to go myself partly for the pleasure of working with those two remarkable young men. But I went to Oak Ridge for a more fundamental reason: I wondered what I could learn from a new laboratory. What novel ideas could I bring to it? And most important, could I help make the fruits of nuclear science available for peaceful ends?

The former head of Clinton Labs, Martin Whitaker, was a physicist my age who had joined the Manhattan Project in 1942 and had been director of Clinton Laboratories since

1943. But he was now leaving to become president of Lehigh University.

The Monsanto Company helped run Clinton Laboratories, and a Monsanto man named Charlie Thomas told me that Clinton Laboratory badly needed me to lead its uranium power development project. There I would join a man named James Lum as co-director of the laboratory. About 400 scientists and technical staff would be under our direction. I accepted the offer.

What a change it was from the Metallurgical Lab. During the war, we always had a goal and a tight schedule. At Clinton Labs, we did not; our exact purpose was unclear. That complicated the work, but brought it far closer to pure science.

The U.S. government had restricted to Los Alamos all the permanent work in the theory and production of atomic bombs. So we worked on nuclear power and nearly forgot about atomic bombs. We decided how to build nuclear reactors and care for them; how to properly introduce their coolant and extract their heat. I liked this bit of practical engineering.

And I used my physics, too. My early work in crystal symmetries had detailed why a symmetric system develops more predictably than a nonsymmetric one. And this knowledge of symmetries greatly shaped my nuclear reactor designs.

The laboratory even did some direct charity work. On August 2, 1946, I stood with one of our engineers in my role as director of research. Together we gave a one-millicurie unit of carbon 14 to a St. Louis hospital for the treatment of human cancer. That was a moment of honor.

But I hated the bureaucracy at Clinton Labs. They called me a leader, and I offered personal ideas on many aspects of nuclear research. I did as well as I could do. But I am afraid that I made a poor boss. Unlike Arthur Compton or Robert

Oppenheimer, I did not easily sense the needs and desires of my subordinates. My chief success came in convincing Alvin Weinberg, Gale Young, and a few others of the value of their own work.

A year at Clinton Labs convinced me that I did not belong there any longer. I had obtained a number of patents in reactor design and I still loved nuclear engineering intensely. But I was never a natural administrator. At heart, I was really a teacher, an old-fashioned scholar who admired small laboratories and independent research.

In the laboratories of large institutions, people get too used to each other. Human thought, which is wonderfully personal, becomes institutionalized. The sense of fresh adventure evaporates. Everything is elaborately contrived; the equipment is quite impressive, but the devotion is gone. Such settings stifle innovation and make me feel useless and sad.

I felt almost like a janitor at Oak Ridge—quite a prestigious janitor, of course, with a spacious office, a range of privileges, and a staff who consulted my opinions assiduously. But still a janitor. Emotionally, the job could not hold me, and I determined to leave it.

I met with the higher-ups at Clinton Labs and told them I had better not remain. They must have known that I was not an effective boss. But they were far too polite to say so. "Say, Wigner, you seem to be doing quite poorly down here." No, that does not sound like them at all.

And since the higher-ups were all acting so politely, I saw no reason to tell them that Clinton Labs had institutionalized human thought. I merely told them that I badly wanted to return to Princeton University. They gracefully allowed me to go without making me feel I had abandoned a commitment.

Alvin Weinberg took over Clinton Labs soon after I left and very slowly improved things there. What patience he had!

Weinberg truly understood bureaucrats and the troubles they faced both from inside and out. He even wrote an incisive book on such things, *Reflections on Big Science.*

So Alvin Weinberg stayed to reform Clinton Laboratories, while his dear friend Eugene Wigner made a quiet retreat to Princeton and the sweeter, more independent life of a theoretical physicist and teacher. But I did not wash my hands of Clinton Labs. In the last 45 years, I have visited Oak Ridge more times than I can count, observing, advising, asking questions, and trying to praise their work.

❧        ❧        ❧

I returned to Princeton University and found it much improved. The Second World War seemed to have left the university far more contemporary and less conceited. And a first-rate university is livelier than a big laboratory. New people are always arriving who know nothing of your academic traditions. You must explain what you do and why you do it to a steady stream of bright students. Some of them will challenge what they are learning. So you are always reexamining the value of your ideas. And that is very healthy.

Working on the atomic bomb had made me better known. The man on the street still knew nothing of me. But if you found a nuclear physicist on the street and asked, "Who is Wigner?" he might well have identified me. A great many physicists were now being invited to sit on various panels, councils, and committees. I joined the throng.

I sat on the visiting committee of the National Bureau of Standards from 1947–1951. I was on a mathematical panel of the National Research Council from 1951–1954, and also on a physics panel of the National Science Foundation at nearly the same time. From 1952–1957 and again from about 1959–

1964, I served on the Atomic Energy Commission's General Advisory Committee. This was a committee of objective outsiders who advised the rulers of the Atomic Energy Commission. We gave earnest advice, and did no harm. But we did not invent anything new.

My own chief bit of advice was to relax the security rules of Atomic Energy Commission laboratories. Clearly, we should not show our enemies how to make nuclear bombs. But I thought that the basic reactor designs should be widely known and that sharing knowledge of nuclear energy broadly would ultimately help our country and our reactor program far more than it would hurt.

Most of the men on the Atomic Energy Commission heartily disagreed with that notion. They did not want the basic reactor designs broadly known. Quite the contrary. They seemed to me unduly motivated by a fear of embarrassment, as though the prevailing political and technical arrangements surrounding nuclear weapons might well be discredited on close inspection. And so we heartily disagreed. But all of this discord passed without rancor.

Far more disturbing were a series of personal tragedies that I witnessed in the 1950s.

In 1954, Robert Oppenheimer, who had served his country so brilliantly during the Second World War, was stripped of his government security clearance. Since 1945, Oppenheimer had been quite politically prominent. He had served as chairman of the Atomic Energy Commission's General Advisory Committee. He had advised presidents on nuclear matters and served as a consultant to the CIA, the Congress, the State Department, and to our United Nations delegation. He was still a superb scientist and had organized several major scientific conferences. Dr. Oppenheimer was a very eminent man and everyone knew it.

Yet by 1954, a fair number of people disliked Oppenheimer, and a few of them wondered publicly if he was reliable and patriotic enough to see classified documents. A hearing was set for April 1954 to settle the matter. About 40 witnesses testified at the hearing, including General Groves, Hans Bethe, Enrico Fermi, Jancsi von Neumann, and Edward Teller. Teller testified essentially that he could not understand Oppenheimer and therefore could not quite trust him. Though Teller was just one of many witnesses, and his testimony by no means entirely unfriendly, it was regarded as among the most damaging to Oppenheimer.

Many people have asked me why Teller testified against Oppenheimer in this case, and if he really hoped to see Oppenheimer's clearance stripped. I cannot say why Teller testified; we have hardly ever discussed the matter. But I think he did so for several reasons.

First, Teller is a man with a strong, patriotic sense of duty. Asked to testify about his country's security, he felt obliged to do so. Second, Teller is a deeply honest man—perhaps too honest for his own good. When his opinion is asked on an important question, he feels compelled to answer fully and from the heart, regardless of whom he may offend. Teller had strong feelings about Oppenheimer, the United States, and the role that nuclear weapons should play in its future. He felt a duty to his country to convey those feelings.

People say that honesty is the best policy and probably it is; but it has certainly brought Edward Teller much criticism. It often seems to me that in America well-known people are tacitly expected to forsake honesty in favor of tact. Teller does not abide by that expectation and he has often been punished for it.

Finally, Teller believes in the value of organizations. He had felt strongly since 1945 that the United States needed a smooth-running division for nuclear planning, whose top leaders ably promoted the majority view on nuclear and military matters. Robert Oppenheimer was not a leader of this kind. Naturally, then, Teller felt uneasy at the prospect of Oppenheimer making crucial decisions about our nuclear forces. And when Teller was asked, he said as much, though it would have been far easier for him to remain silent.

Behind Teller's words at this security hearing was much of the old frustration we had both felt with Oppenheimer before the war. Teller must have remembered what Oppenheimer had told us then: that Hitler was hardly dangerous and would surely be defeated before doing much harm; that higher-ups would see to all this; it was not our problem.

Oppenheimer had really said those things before the war, and I had been as angry with him as was Teller. But I had completely forgiven Oppenheimer after his brilliant work at Los Alamos. Teller could not quite do so.

One man forgives; another does not. Who knows why? I wish I understood better the mysteries of human character. Though Teller's dislike of Oppenheimer at times seemed a bit extreme, I never saw any sign of prejudice against him and never pressed the issue.

One thing I do know: If Edward Teller could have forgiven Oppenheimer, he would gladly have done so. Edward does not enjoy holding grudges. He simply could not forget that Oppenheimer had once been insufficiently devoted to the military defense of freedom.

Robert Oppenheimer certainly should not have lost his security clearance. He would never have betrayed the United States, and Teller himself said so during the hearing.

But the hearing board *did* take Oppenheimer's clearance from him. And this disheartened Oppenheimer quite seriously. Of course, a man of his eminence does not disappear just by losing his security clearance. Oppenheimer was made director of the Institute for Advanced Study in Princeton and given a handsome salary. He was consulted a good deal by journalists and historians as well as scholars.

But I doubt that Oppenheimer's last years were nearly as satisfying as they should have been. He lived until 1967, and I often used to see him out walking around Princeton. He did not look happy. He had been disgraced, and I think he felt that very deeply. The loss of Oppenheimer's security clearance gained Edward Teller a great many new critics. I was not among them.

In 1955, came another day of great sadness: the death of Albert Einstein. Einstein had been the most famous scientist of our century and also the greatest, which are not often the same thing. Yet Einstein had never said, "I am Albert Einstein! I have studied physics longer than you, and with far greater insight!" He had never wanted to float up in the sky somewhere, above the rest. Einstein spoke of science objectively, without referring to his own success. That was a great lesson to me.

Around 1950, Einstein told me, "I am not in good condition to answer complex physical questions. I am getting older. In fact, I am getting old." And so he was. Like many scientists at that stage, Einstein became largely a philosopher. He was still enormously friendly and imaginative, still deeply interested in philosophy and politics. He continued to help other physicists and to deserve all the admiration of his peers. But he was no longer changing physics. Happily, he was as profound a man as he was a physicist.

Einstein's deep interest in politics was less in electoral results than in the psychology underlying all human political activity and the philosophy to be adduced from it. I sensed that Einstein did not fully trust American democracy and the quality of public thought in this country. I wish now that I had explored this topic with him further.

Einstein strongly approved of the creation of Israel. He and I agreed that Israel's existence was very good, at least for the Israelis. We spoke of visiting there. In 1952, Einstein was offered the presidency of Israel, but I doubt that he ever seriously considered accepting. Besides his advanced age and complete lack of administrative experience, I think Einstein did not want to live in Israel or become an Israeli citizen.

Einstein was quite fond of the Jewish people, but he rejected all religious dogma. He felt that religion belonged mainly in the ethical domain, its historical and scientific assertions kept in the background.

Sad to say, Einstein hardly enjoyed his last years. The extinction of life was a subject he had thought and written a good deal about. But like most men he could not enjoy the last years of his own decline. He died in a kind of solitude. My only comfort at Einstein's passing was my strong impression that he had found death a liberation from obligations.

Then in 1957 something else terrible became clear to me: that my dear friend Jancsi von Neumann was dying of cancer. I had known Jancsi all my life: as a child in Budapest, as a young man in Berlin, and finally as an adult in Princeton. Jancsi had been affiliated with the Institute for Advanced Study since 1933. He had married again in 1939—an old Hungarian friend, the former Klari Dan. He had enjoyed life immensely.

I went to see him several times near the end. He was not his cheery self and I tried talking to him. We all know this type of conversation: I asked him what I could do for him. Then with a burst of false cheer, I forecast that his illness would recede, that he would regain his strength and return to his work. But I knew that Jancsi was incurably ill and I made a poor actor.

And Jancsi knew it too; his brilliant brain was realistic to the end. It told him plainly that he would soon stop thinking, that the man who had been Jancsi von Neumann would soon entirely cease to exist. Yet I think the full meaning of this idea was incomprehensible even to Jancsi, and what he could understand of it he found horrible.

Von Neumann's life had been a long series of brilliant achievements. He had fulfilled all of the astounding promise that Rátz had seen in him 45 years before. He had made himself one of the great mathematicians of our century. There were just two mathematical fields that von Neumann had not advanced: topology and number theory, and he had plainly mastered number theory.

Von Neumann had done crucial mathematical work in group theory, particularly with noncompact groups. In mathematical logic, his work prefigured the great work of Kurt Gödel. His theory of self-adjoint operators was a magnificent expansion into the complex and infinite dimensions of Hilbert space.

Jancsi had also done superb work in game theory. That an optimum method exists in playing games of chance does not surprise even an amateur card player. But the mathematical basis for this is surprisingly difficult to establish. Von Neumann did it.

Jancsi had failed just once in his mathematical career: He had been unable to solve a prime question of mathematical logic with the tools of Hilbert's axiomatic school. He had admitted his failure and given up. When the right solution was found by others, Jancsi saw it at once, and admired it sincerely.

Jancsi was not only brilliant, but unfailingly kind to his

fellows. He knew that guidance of one's peers is best done casually. He may have advised more work than any other modern mathematician. His own papers had an elegant power and rarely needed revision even years after.

Then there was his conception of the first modern computer. The theory and practice of computers was his chief mental love at the end. Jancsi was at the head of a group that first designed and then built the computer capable of using a flexible stored program. With typical wit, Jancsi named this computer MANIAC I.

Many of us failed to see its full value for years after, but Jancsi saw that computers could help solve problems in mathematics, physics, economics, industry, and even the military. He wrote part of a series of papers on the theory of logical organization by computer. His work on computer coding and programming is seminal in that field.

The computer made a graceful cap to von Neumann's career. Ever since he was a schoolboy, Jancsi had served as a kind of computer to his friends, always ready to make complex calculations. He had always loved axioms, and it was wonderful to see him writing axioms in computer language in his last years.

Jancsi spoke five languages and also read Latin and Greek. He had done important work in economics: the role of profits, the benefits of competition, and the best methods of developing national economies. He read widely and remembered it all, whether technical or literary. Large chunks of world history he knew thoroughly, especially Greek history. The history of the Renaissance he knew as thoroughly as a professional historian.

Von Neumann had also served his adopted country in warfare. During the Second World War, he had been a valued consultant to the Army, Navy, and Manhattan Project. Working independently, he had nearly discovered the implosion method of bringing nuclear fuel to explosion. His work on the lens principle of detonation was invaluable. In 1954, von Neumann had been appointed to the Atomic Energy Commission. The chairman of that commission said that once Jancsi had analyzed a question, no further discussion was needed.

That is a great deal of success for one man, especially one who calls himself merely a mathematician. And despite the variety, one saw that all of Jancsi's achievements rose from a common point of view about life. Whether eating imported sweets, sleeping late in the morning, solving a protracted mathematical puzzle, or driving his Cadillac, Jancsi lived in elegant style. He had lived a coherent life as well as an honorable one.

Yet all of this glory could not reconcile Jancsi to his destiny. It was terrible to watch his frustration when all his hope was gone. What a contrast he was to Enrico Fermi, who was so composed by death's approach he seemed superhuman. Ten days before Fermi passed away, he had told me: "I hope it won't take long." He had reconciled himself perfectly to his fate.

But Jancsi could never to that. His agony was appalling. Jancsi von Neumann died on Feburary 8, 1957. He was just 53 years old.

Every few years after Jancsi's death came news of the death of another mentor or colleague: Wolfgang Pauli in 1958; Max von Laue in 1960; Niels Bohr in 1962.

⚜        ⚜        ⚜

I traveled a good deal in the late 1950s and learned much about the world.

In the summer of 1959, I was one of four principal lecturers in a six-week conference on the principles of invariance. The conference was conceived by the Latin American School of Physics and held in Mexico City. Its 35 student participants were from all over the Americas.

The conference brimmed with cordial hospitality. Lectures were held back-to-back on Tuesdays, Thursdays, and Fridays. Absorbing them all, each with its own discussion period, was exhausting. But it left us with long, leisurely weekends to see Mexico.

I was enchanted by Mexico's color and natural beauty. It has more archaeological relics than any country I had ever seen. To live 10,000 feet above the ground was quite invigorating. And just beyond this bustling modern city were great mountains and volcanoes, forests and brooks, an ancient way of life. I was charmed.

In July 1960, my wife and I saw Israel for the first time. It was a vivid period in the life of that young country. Just a few months before, Israel had announced the capture of Adolf Eichmann in Argentina. Germany was still paying war reparations.

Israel was developing rapidly, a nation clearly serious about its future and aware of the deep obstacles in its way. New buildings and industries were everywhere. Standards in science and the arts were high and still rising. Israel's biggest troubles seemed to me the role of the ultrareligious groups, the position of the Arabs, and the love of intrigue that seems to afflict all small nations.

I spent some time at a kibbutz and greatly admired the voluntary socialist ethic of the kibbutzim. But the kibbutzim youth surprised me. They seemed sure that their cooperative life-style would in time become nearly universal. And they seemed to believe that history began in 1948, with the end of the British mandate in Palestine, proclamation of the state of Israel, and its attack by Arab armies. These kids hardly cared about anything that had occurred before 1948. So it often is with children.

�֍     ✖     ✖

A curious thing happened to me in December 1963. On one of my periodic visits to Oak Ridge National Laboratory in Tennessee, a man I barely recognized came quickly over, grasped my hand heartily, and shook it.

"Dr. Wigner!" he burst out. "Let me congratulate you!" I looked back at him, thinking how pleasant it is to be congratulated. After a moment I said, "Thank you very much! But why are you congratulating me?" The man said, "Well, you have received the Nobel prize!" I briefly considered the meaning of this statement. "No!" I said firmly. "You must be mistaken."

I had never, ever thought of winning the Nobel prize. That is really true. I had grown up never expecting to get my name in the newspapers without doing something wicked. So I could not believe at first that I had been awarded the Nobel prize.

But in the next few hours five or six other people came over to me and insisted on clasping my hand in congratulations for the same reason. So I decided that it must be true. After all, one man can make a mistake. But it is not plausible that six men will make the same mistake in short order. And at home I found a piece of mail confirming it: I had won the Nobel prize in physics, the highest award a physicist can earn.

Winning was a great pleasure. But I doubted that I had achieved enough to deserve a Nobel prize. I thought of the old German proverb: "The stupid one has luck."

Einstein, Max Planck, Max von Laue—these were men who had clearly earned the Nobel. The deeds of men like Ernest Lawrence and Werner Heisenberg were dramatic, and we fully expected the Nobel prize for them. In Dr. Wigner's case, no. I had done some fine work, but much of its power was latent. I had never invented anything so clearly fundamental as Lawrence's cyclotron or devised a principle as fundamental as Heisenberg's quantum mechanics.

They give a Nobel prize in physics each year, and in 1963 they split the prize among three scientists. So at first I thought, "Perhaps they have finally exhausted the supply of worthies." But a few minutes' thought refuted that idea, for among the scientists who had never won the Nobel prize were Jancsi von Neumann, Robert Oppenheimer, and Edward Teller.

Now, von Neumann was known mostly for his work in mathematics; Alfred Nobel did not sponsor a mathematics prize so I was not really surprised that von Neumann had never won a Nobel. But I felt that Teller and Oppenheimer certainly should have won the Nobel before I did. Many other physicists should have.

I thought the prize might even have come partly as a reward for my political acts, principally for the work that I had done 25 years earlier in preparing the United States to resist the Nazis. But I hoped that this was not the case, and I never advertised my fear that it was.

Collecting the prize was a great pleasure. I arrived very safely in Stockholm and was met with kindness. An elaborate Nobel prize ceremony was staged in the Stockholm Concert Hall. More than 1000 people attended with journalists from many countries. Nobel prizes were awarded in chemistry, literature, and medicine as well as in physics.

A short introduction was read, formally stating why I had won the prize: "For systematically improving and extending the methods of quantum mechanics and applying them widely." Then the King of Sweden presented me with my Nobel prize.

Every Nobel laureate is expected to make an acceptance speech. I had also written a technical paper for the occasion, and even learned a few words of Swedish.

There followed a Nobel Foundation banquet at the Stockholm Town Hall. I was reintroduced to the King of Sweden. He served all of the Nobel prize winners of 1963 a hearty meal, and we all smiled at him and praised the temperate climate of Stockholm. The king was hardly flattered. He already knew that Stockholm's climate was temperate and had heard it toasted many times before.

I was sorry my parents had died without knowing that their son had received a Nobel prize. But I was able to share the prize with my colleagues, my wife, sisters, and children.

The Nobel prize doubled my mail. People like to contact Nobel laureates with questions and suggestions. Many of them also asked me to send money. But, fortunately, people did not seem to need to know my personality, as they had needed to know Einstein's. And I felt no greater pressure from my Nobel prize. I knew it was a reward for past work, not a challenge to begin something novel.

Winning the 1963 Nobel had just one distasteful result: It briefly strained my relations with Leo Szilard. Szilard made clear his feeling that before Eugene Wigner won any Nobel prize, Leo Szilard should have won it. Szilard was hardly undecorated with awards. He had won the Einstein award in

1958, and he and I together had won an Atoms for Peace award in 1959. But Szilard had never won the Nobel prize, and he felt that his own deeds had been ignored—not so much by scientists as by the general public.

Szilard had turned to biology in 1949, and it was in that field that he came closest to realizing his full potential. He had stayed at the University of Chicago after the war. I never knew his exact position there, but he seemed to have all of the respect and freedom due to a senior professor.

Szilard had found a strong younger colleague, Aaron Novick, to aid in his researches. Szilard had hatched some fine ideas about bacterial mutation and biochemical mechanisms and created some bacteria-studying devices. While Novick tried to execute these ideas, Szilard's eager imagination was free to roam new regions.

Szilard was very interested in causality: whether the world can be predicted. He believed deeply that it could, and being Szilard, he argued that point strongly. Causality was a fine idea, but Szilard felt that he would never be able to calculate causality and he hardly tried to do so. I found that sad.

Szilard never formally left the University of Chicago. But he often sought other positions. His long pursuit of jobs diverted him further from science. I think he could have thrived as a corporate consultant or as an employee of a publishing house. But he did not seek these jobs with any consistency or tact.

Universities that might have hired him feared and exaggerated his selfishness. A few jobs were almost offered, but by demanding them, Szilard lost them. I had tried hard to get him a teaching position at Princeton around 1938, but not as a top professor. And Szilard would very likely have turned it down.

I was often peeved when Szilard changed plans without warning. He was always turning in a new direction. His friends learned to accept that. I had often disagreed with Szilard, but we had a good idea when discussion would be futile, and until my receiving the Nobel prize we had largely avoided it.

Szilard knew his wife, Trude Szilard, for over 20 years before marrying her in 1951. Even then, they did not live together for another ten years or so after the marriage. So even Szilard's best friends were often confused as to whether he was married. But it became increasingly clear as Szilard aged how much he appreciated Trude. She brought him steadiness and order. I have already noted that Szilard always feared becoming insignificant, or even appearing so. Many adults rely on a loyal spouse to believe deeply in their significance. Szilard relied on Trude very strongly in that way.

And Trude handled Szilard beautifully. She saw him a great deal, loved him a great deal, and assured him that he was truly important. And that made a very great difference to Szilard. I think without Trude, some physical or emotional catastrophe would eventually have befallen him.

I gave Szilard money near the end of his life, as did a few other of his friends. Szilard was nearly poor then, and he wrote us sincere letters of gratitude. He was in the hospital a good deal in 1959 and 1960. I feared that he might be too ill to receive in person our joint Atoms for Peace award, in May 1960, at the National Academy of Sciences in Washington, D.C. But he was there, both in body and spirit.

Leo Szilard was about 66 years old when he died on May 30, 1964. I had known that he was not well; still, I was deeply shocked at the news. I had not prepared myself emotionally for his passing.

In his last years, Szilard and I had increasingly disagreed on important questions. But the friendship of a lifetime is not erased by disagreements of a few years' standing. Szilard was my friend, an advisor and companion over many years and on two continents. We had learned much science together, enjoyed and survived much together. It was very hard for me to accept that he was gone.

＊     ＊     ＊

Szilard's career clearly had less scientific value than that of von Neumann, Einstein, or any of the greatest scientists. Szilard had never given science any great new idea. If Szilard had been able to peacefully live out his life in Budapest or Berlin, he would likely have been celebrated for the breadth and savor of his ideas alone. But in the United States, where ideas themselves were less highly revered, scientists were expected to refine their ideas with careful calculations. And Szilard was not this kind of scientist.

But let us look at what he did achieve: Szilard's work had graced statistical mechanics, nuclear physics, nuclear engineering, genetics, and molecular biology. He had invented a method for pumping liquid metal. He had founded at least two associations to shape the policies of his adopted United States. He had even written a charming novel, which was translated into six languages. And Szilard is the man who can most plausibly be called the "Initiator of the Manhattan Project."

Szilard was wrong about a great many things, but he was right to feel that he had never received his scientific due. The popular ignorance about Leo Szilard has always pained me. It is one of the chief reasons that I decided to leave the world this book.

Szilard had weaknesses, yes—quite irritating ones. No one knew them better than I did. But every one of his weaknesses was essentially harmless. For all of his faults, Leo Szilard was the best friend I ever had.

# Seventeen

# *The Gold That You Have Will Finally Kill You*

*L*et me tell you a fable:

Once, long ago, there lived a rich and powerful caliph. He ruled his domain for many years, justly and in peace. His subjects were prosperous, indolent, and happy. But one day Mongol tribes heard of this felicitous land and began plotting to overrun it. They raised an impressive warrior band, set out for the caliph's land, and were soon knocking at its gates.

This great caliph had just one fault: He could not bring himself to spend any portion of his riches, least of all on armies and fortifications. The inevitable came to pass. The Mongols broke down the feeble gates protecting this peaceful land. They invaded, sacked, and pillaged at will. When they captured the royal palace, they found the caliph sitting in his strong room, amidst his treasure hoard.

The Mongol Khan was a brutal man. He was not content

to hold the caliph for ransom. He wanted him executed, and in a quite unusual way: molten gold poured down the throat. "You see," said the Khan. "That gold which you have hoarded so assiduously will finally kill you." And so it did.

Now, since the end of World War II, the United States has not been an Eastern caliphate. But it has been a very wealthy nation. And we have too often acted like that foolish caliph. We have built a mighty arsenal of airplanes and battleships and nuclear bombs. But we have largely ignored civil defense. And this has needlessly exposed our nation to danger from the Soviet Union and, increasingly, from other nations as well.

"Civil defense" is a general term, covering a variety of ideas designed to save lives in a nuclear war. There are two major areas of civil defense: first, the evacuation of major cities; and second, the protection of people in heavy, well-protected bomb shelters.

Defensive measures both in warfare and diplomacy have nearly always been grossly underestimated. If we truly understood the human mind, we might learn why. But this has clearly been the case.

Before the First World War people dismissed the value of trenches. On the eve of the Second World War foxholes were not valued. And I feel sure that if we ever have a large nuclear war, we will find that bomb shelters too have been quite underestimated.

During the Second World War, I felt that large-scale civil defense should have begun. But I shared this view with just a few colleagues and friends. I never tried to study the issue technically or to make a political case for my views, as I had done a few years before with the atomic bomb. After the war, civil defense was for a time hotly debated and I publicly backed the idea.

I saw the United States investing huge resources to keep a strong military. We believed that such force deterred foreign aggressors, and probably it did. But after 1952, when the Russians conclusively proved that they too had atomic bombs, the world seemed to me in intolerable danger. Offensive weapons alone could not guarantee our safety.

The official policy of the United States was then based on something called "mutual assured destruction." Perhaps it still is. The idea of mutual assured destruction was that the Soviet Union and other nations would have to be crazy to attack the United States or its allies.

If the Soviets did attack us, the theory held, we might well respond with a nuclear attack. They would then have to launch their own nuclear bombs, thus "mutually assuring" destruction and large-scale human death. World peace, then, depended largely on the revulsion our enemies were presumed to feel toward massive devastation.

But the mutual assured destruction policy rested on a pair of mistaken beliefs: that enemies are rational in wartime, and that other cultures value individual human life as Americans do. In fact, many do not. I suspected then, and I still suspect today, that either Joseph Stalin or Nikita Khrushchev would have been willing to lose two thirds of Russia's people in order to control the globe. Happily, such mass sacrifice is not a real option for an American president.

But I used to worry about our president getting a sudden phone call from Russia or some other country. A foreign voice would come on the line: "Unless you allow us to station military regiments in Buffalo, New York or St. Louis, Missouri, tomorrow evening we will destroy all of your big cities. Most of your people will perish." Or perhaps they might insist "only" that we abandon South Korea to communism.

Now most U.S. citizens would prefer to abandon South Korea to communism than to see millions of Americans die. South Korea is a small and distant land. The average American scarcely notices it. Communist rule is rather abstract to most Americans while the fear of nuclear attack is frighteningly real.

So I understood the instinct to accommodate the nuclear terrorist in order to forestall a nuclear strike. But accepting the terms of a blackmailer only invites further threats.

Before long, I felt, a nation must reject the blackmailer's terms and prepare to fight some kind of nuclear war. And I was convinced that once the false security of mutual assured destruction was stripped away, Americans would wonder why their leaders had spent so little time preparing a common defense against nuclear attacks.

In the 1950s and 1960s, I tried to expose the foolishness of mutual assured destruction. I was disturbed that civil defense did not much interest the average American. For one thing, it was presented poorly. A rash of novels and films after 1945 depicted the horrors of nuclear war and the "futility" of civil defense. It was easy to see why these books and films were popular. Nuclear war makes a fine subject for a horror story. I might have read such books myself if I liked the horror genre. But such products should never have been treated as rational guides to anything. And yet they were. It was years before I fully understood how largely public opinion is shaped by popular entertainment.

But the fault was not only with popular culture. Even "factual" books on civil defense written by scientists vastly simplified its problems. Scientists are trained to think clearly, and when they do, they can clarify and sharpen political debate, which is too often governed by ignorance, clever lies, and the exploitation of crude emotions.

But most of the books that scientists wrote about civil defense replaced logical analysis with overt emphasis on certain political values. Their authors presented these books as rational guides, but what they resembled most were the tracts of ideologues.

I heard a great many arguments against civil defense in the early 1960s. Nearly all of them could have been applied generally to all defense measures. I mention here some of the most common.

First came the argument that since nuclear war was unlikely, the United States could not afford extensive and costly precautions against it, with so many vexing social problems already in our midst. To this, I responded: It is rarely easy to know how near we are to war. And it is human nature to deny the likelihood of any great calamity. But let us suppose that nuclear war is unlikely. Suppose that over many years a huge sum of money is spent on a defense that is never used. Should we feel cheated? No! Civil defense is a kind of national insurance. Do we feel cheated after 20 years of fire insurance when our home remains uncharred?

Another popular argument was that civil defense could never save very many lives. People insisted that any survivors of a nuclear war would perish anyway from nuclear radiation or from economic dislocation. There was in the 1960s a very popular phrase about life after a nuclear war: "The living will envy the dead."

How that phrase irritated me! If you mentioned nuclear war, people said, "Yes, you know after a nuclear war, the living will envy the dead." Well, in Europe after the First World War, the living did not envy the dead. In Eastern Europe during and after the Second World War, very few of the living envied the dead. No society lost its desire for life despite a total economic collapse and the ruthless cruelty of the Soviet Union.

Before the Second World War, British government circles felt that if British cities were subjected to heavy air raids, many of their civilians would give up and roam the streets hopelessly. Yet British air raid shelters proved to be quite effective in that war, both emotionally and physically. So did the German bunkers.

People are strong. Societies do not die easily. Statements like "the living will envy the dead" are not rational arguments about anything. They do not suggest how we might avoid a nuclear war or refute the value of civil defense. But they are laden with a sense of doom that people find very attractive.

I was convinced that no matter how a nuclear war began, the United States would have a duty to emerge from it as a viable nation. We would have to be able to recover our social order and most of our economic power within 20 years. This would be far easier with well stockpiled supplies, at the very least such staples as food, medicines, tools, and gasoline. Without such provisions, I was sure that the survivors would suffer needlessly. They would have a perfect right to feel betrayed by a political leadership that had failed to prepare the nation.

Other opponents of civil defense insisted that we should prefer to "die on our feet" than to "live on our knees," in an underground bomb shelter. They called it humiliating and cowardly to crawl into a hole in the ground to avoid nuclear attack. I found this argument curious indeed. Would it be more becoming to stand in the open and be slaughtered? Man has always sought shelter from his enemies.

The symbol of America in the war against Hitler was the

young soldier in a foxhole. Was it cowardly for him to refuse to meet enemy shells in the open? No! True courage is not an eagerness to die, but a willingness to face danger squarely. It is courage that puts the soldier in a foxhole and courage that prepares bomb shelters for our civilians.

Some people argued that any nation preoccupied with civil defense was dangerously neurotic. They implied that if our leaders were obliged to think too much about nuclear war they might become nervous and irritable, anxious to "get it over with."

Again, I could not agree. The danger of nuclear war is real and constant. To see that is not to have a neurosis. Hospitals arouse neurotic feelings in us, but we spend a great deal of money on hospitals. Cemeteries arouse neurotic feelings, but we maintain cemeteries. They serve a crucial purpose.

I found myself asking, "Which is more neurotic: to make careful plans in case of nuclear war or to largely deny the chance of such a war?" A strong civil defense program should certainly include educational programs about the horror of a nuclear war. A graphic education of this kind would hardly produce leaders anxious to "get it over with."

I was amazed by some of the reasons given against civil defense. At least one man argued that bomb shelters could not work because radiation would inevitably seep into the shelters and destroy the electric motors that drove the shelter fans.

Some people objected to the prospect of admitting unpleasant neighbors into crowded underground shelters for long periods. All over the country, people debated the extent of the moral obligation to accept newcomers in such a shelter.

This at least was a real moral issue, like that of a man asked to share his last loaf of bread with a neighbor. But I could not help thinking: Why not work to get everyone a loaf rather than refusing the first man his loaf?

Still others resisted civil defense on the grounds that it would do nothing to bring peace. That objection was entirely true; civil defense would *not* bring peace. But so what? Many important national acts do not promote peace. Building the interstate highway system does not bring peace. The first duty of a nation is to its own security. Only then can it properly pursue peace. And from a moral point of view, I felt that protecting our own people with defensive measures was far better than designing more and greater weapons with which to kill foreign civilians.

The most serious critics of civil defense granted its utility. But they argued that it must be judged principally by whether it made nuclear war more or less likely. And since most of them felt that civil defense made nuclear war more likely, they condemned it. I considered this argument very carefully. But again, I could not agree.

Since the Second World War, both we and the Soviets have had many more offensive weapons than defensive. Such imbalances tend to make a first strike attractive to war planners, especially if they see war as inevitable. But if the Soviets could not be assured of victory, I felt they were unlikely to attack. The Soviets tend to be reckless once war has begun, but they have been cautious about starting wars. So I felt we should put in balance our offensive and defensive forces.

Making major reductions in the nuclear weapons of either power was politically impossible after the war. Civil defense seemed the best alternative. It required neither good diplomatic relations with Russia nor on-site inspections of their weapons facilities. It also promised healthy insurance against the threats of bomb-wielding countries other than Russia. I felt sure that the number of countries with nuclear weapons would steadily grow.

I began to feel that the incongruity of arguments against

civil defense had an unadmitted origin. People would pay taxes to support an army, but they did not want to hear about actual warfare, wholesale death and destruction. They wanted to live in a refined atmosphere into which brutality did not penetrate.

This, too, is human nature, but it is unhealthy. No nation has ever survived very long whose citizens considered military defense beneath their dignity.

The civil defense debate again brought home the lesson my father had told me when I was a youth, brimming with ideals: People do not build their beliefs on a foundation of reason. They begin with certain beliefs, then find reasons to justify them.

I gave as many talks about civil defense as I could. President Kennedy and a few aides even listened politely to me for several minutes in the White House.

In the summer of 1963, the National Academy of Sciences made me director of the Harbor Project, a six-week study of civil defense conducted by 62 scientists, engineers, and statesmen. Our goal was simply to study the issue of civil defense, without political bias. We were certainly not expected to recommend the adoption of civil defense measures. But by the end of the study, the use of nuclear blast shelters seemed to us so clearly sensible that our report became, in part, a political document urging their construction. I felt that a strong civil defense program could be maintained for about one tenth of the total defense budget.

By 1968, civil defense was no longer a popular subject. Everyone was talking about Vietnam. Subtler political, military, and economic issues were nearly ignored. Politics is often that way; vital issues persist for decades, but the seriousness with which they are treated shifts rather drastically.

I kept trying to arouse my colleagues about civil defense.

But when I made my case, most of them said, "Oh, Eugene. Not again. I am a scientist, not a politician. Please leave me alone." My work for civil defense was not popular among the scientists I knew.

And civil defense has never returned as a major political issue. I am afraid that our neglect of it may prove disastrous. The Russians maintain a strong civil defense program, though they have less need of it than we do because our leaders are more peaceful and our cities more densely populated.

The Russians have put civil defense programs in every grade school and high school. They have made factory directors responsible for it in major factories. Their civil defense handbooks boast of evacuating large areas in 36 hours. Elaborate evacuation plans have been made and practice runs announced over radio and television and in the newspapers. Transportation points have been specified; baggage limits set; food, lodging, mail, and medical services planned for, even special transportation for the sick and infirm.

Plainly, the Russians could not protect every citizen. Perhaps 5 million Soviets might die in a major nuclear attack. That is a serious thing; you could say that the United States had 5 million Soviet hostages. But then the Soviet Union had 80 million American hostages. And however terrible 5 million deaths would be, they would make up less than half as many people as the Russians lost in the Second World War.

Many people do not realize that bomb shelters also make good refuge from earthquakes or floods. Some of the subways in Moscow, Leningrad, and Kiev are nearly 120 feet underground, with blast-proof doors 18 inches thick. Their many miles of subway tunnels protect against nuclear fallout and nuclear blast far better than most subway tunnels in the United States.

The Swiss, who take pride in avoiding wars, built fine bomb shelters in the 1960s. In Switzerland and Sweden, nearly every new home erected then was given a built-in bomb shelter. China has also practiced good civil defense.

Meanwhile, the United States has had very few bomb shelters, few detailed evacuation plans, and very little practice executing them. Most of our fallout shelters have been built in cities, where they would be exposed to blast destruction. So you can see why the lack of a comprehensive civil defense program deeply concerned me in the 1960s. It worries me still.

∞ *Eighteen* ∞

# *The Subtle Pleasure of Forgetting*

*T*he late 1960s were trying years for me and for our country. Most young people in the United States seemed deeply restless. Many of them were ingesting powerful hallucinogens. Much of daily conversation was political, and people of all ages seemed highly agitated.

We were waging a disturbing war in Vietnam. I had no illusions about the costs of war, and I had hoped that we could avoid fighting this one. But President Johnson had called it a duty and had sworn that the war would resist the rise of communism in Asia. For my outspoken support of the Vietnam war, a few people praised me. But it seemed to me that many more people were critical. Rebukes from fellow scientists I felt deeply.

As people age, they tend to see recurring many of the great themes of their youth. On a very deep level, they recognize the reemergence of things from the world of their childhood, and they are astonished to see these things widely hailed as new. Nearly every new trend that old people see in politics and culture reminds them of something else that has come and gone before. This is foolishness; it is also wisdom.

By 1968, I felt much as I had in 1935, when so little stood between Adolf Hitler and Nazi domination of the globe. In 1968, I feared we might have to endure another holocaust to preserve what we call freedom.

I was not always somber. Always I had physics and the company of my family. We often relaxed at a cottage on Lake Elmore in Vermont. The local people were quite friendly. Summer is brisk in Vermont, and we swam and walked the perimeter of the lake. I brought books up to the cottage, but only books with pleasant themes. Life at Lake Elmore had no official character at all; that was its beauty. We felt there a rare and lovely sense of ease.

But even when I was physically content, emotionally I was not. I felt that I was reliving 1933 in Germany, watching ideals that I held sacred flouted by a vocal, self-proclaimed "progressive" part of the nation. And despite the political nature of discussion, people seemed strangely passive to what they saw occurring around them. That reminded me quite powerfully of the Nazi period.

Much of American youth no longer deferred to their elders; they called them old fogeys and held them in contempt. The young seemed to scorn the sacrifices made for them by the U.S. military and failed to see how much our army had done to rid the world of dictators. Many of the young held the military in contempt; some even denounced the scientific world and the "corrupt bargains" that scientists had struck with the military.

And yet, as a rule, these young rebels were quite articulate, every bit as confident of being "the wave of the future" as the Nazis had been 30 years before. As I heard the rebels boast "the future belongs to us!" I knew for the first time why Germany had been unable to resist the Nazis. I had already heard and read scholarly explanations of the Nazi rise. But this I could see and hear all around me. This I felt in my bones.

My own children never joined the rebellious youth— quite the contrary. And it was clear to me even in 1968 that many people of all ages had retained their deep ties to American national ideals. But many of my friends and colleagues chose to rebel against some of the deepest ideals of the country. Angry protest seemed to be everywhere.

The rebellious "wave of the future" had no political program as evil as Adolf Hitler's. Friends told me to take comfort in that. But the rebellious youth did have opinions, and they seemed to me very sick ones.

My friends in 1935 had never accepted in their hearts a single word of Hitler's political program. They had not wanted to resist Hitler by force of arms, but they had never lost sight of the sickness of his views. But by 1968, many Americans whom I respected had quite lost sight of how sick the views of the rebellious youth were. I feared for some years that this rebellious youth might even try to eliminate all the "old fogeys" somehow and to seize power by force.

Had Americans exaggerated the strength and beauty of freedom and democracy? Can democracy function in a land where most people barely discuss political questions? Is there enough sense of purpose and adventure in a democracy to inspire its citizens to defend it, with their lives if necessary?

I wondered if America might one day have to keep order by governing its people with a strictness unknown on these shores. And I felt with a heartsick regret that perhaps the wave of material plenty, the product of that great scientific movement of which I had always been so proud, had deprived our youth of much of the struggle on which man thrives and on which civilizations depend.

I wrote long, gloomy letters to Michael Polanyi in England on these themes, hardly knowing why I wrote. I would not have felt honest without sharing my fears with my great mentor. But I was tempted to close each letter with an appeal not to respond. Such were the late 1960s for me.

I felt that many people, especially in the Democratic party, were loath to defend our country. Most of my friends, even those who supported my views, smiled at the depth of my apprehensions. "Eugene," they said, "surely you exaggerate the problem. These are the United States—the greatest democracy in world history! Why do you speak of dictatorship?"

Well, it is true that nearly all Americans oppose dictators in theory. But life is not lived in theory. Most people do not expect a dictator to arise. When he presents himself, most people do not mind him much if they can ingratiate themselves with his circle or if they feel that he takes their side and only suppresses those whom they dislike.

That is a sad fact of human nature, much denied in America, but nonetheless true. Having learned this truth quite painfully in Europe, neither Edward Teller nor I saw any good reason why Americans should deny it.

Meanwhile, the United States Defense Department was constantly vilified, its budget always in danger of reduction. Around 1972, for example, the Federation of American Scientists attacked a man named John S. Foster, director of research and engineering at the Defense Department.

Among his critics were men I knew and deeply respected, both as men and as scientists. So I struggled to grasp the meaning of their criticism. I concluded that most scientists reject new defense measures unless a specific enemy threat can be conclusively proven.

John Foster and the Defense Department felt as I did: that Soviet secrecy made hard proofs impossible; that our Defense Department was obliged to respond vigorously to potential threats as well as real ones; that nations cannot afford to risk their security. It is very hard to insure the military strength of a great nation, but I saw it as a duty. In Europe, I had seen what happens to countries that cannot resist aggressors. And so I saw men like John Foster as patriots. I hated to see scientists attack them.

*  *  *

My Princeton job expired in June of 1971. I had committed the crime of becoming 68 years old. Princeton, like most universities, asks you to retire from active teaching at that age. And they asked me very gently and politely, as if they were fond of me for having managed to reach such an advanced age. So I retired.

The idea of retirement I accepted freely. My life had been so full of conflicting duties that it had often not seemed my own. Retirement brought some relief. Princeton had made me a "professor emeritus," a kind of professor who retains most of the scholar's privileges and few of his obligations.

My wife and I had more chance to travel in the first months after my retirement. We traveled a great deal. That was fascinating, but no less tiring than life as a professor.

So I was receptive when a man named Roger Richardson, the dean of the Engineering College at Louisiana State University, asked me to come down to Baton Rouge and teach. I had once advised the Esso Corporation, and Roger Richardson had been the Esso employee paid to comfort their consultants. Apparently he had liked me then and still did. He did not even seem to mind that I was 68 years old.

His invitation to Louisiana State would allow my wife and me to see a new university and a charming new region of

the country. I recalled how strikingly beautiful I had found
Wisconsin on arriving there from Princeton 35 years before.

So I accepted the offer, and my wife and I went down to
Louisiana. We were never quite at home in Baton Rouge, but
the people were very friendly. The countryside nearby was
warm, rainy, and lush. And the food was superb. Even a nuclear physicist knows when steamed fish has been well
cooked.

I taught graduate students at Louisiana State. I tried to
show them the beauty of the basic physical theories and then
of the philosophical grounding of physics: that it examines
inanimate nature; that it has laws of nature, initial conditions,
and so on. I wanted them to regard physics as something
lovely, and I think many of them did.

I was pleased when students and even some of the faculty
came to my office to discuss specific aspects of their work.
They received me with more deference than my views deserved, nodded when I spoke, and smiled at me constantly.
For all I knew, they were fine physicists too. But good manners are far easier to gauge than good physics.

When I lectured on thermodynamics and statistical mechanics in Baton Rouge, I was fondly reminded of Einstein,
Szilard, von Laue, Polanyi, and the rest. My contact with the
physics faculty was less close than I had hoped, but the beauty
of the surroundings was just compensation. My wife and I left
Louisiana after just one year and returned to Princeton in
May 1972.

❧        ❧        ❧

In 1974 came a terrible blow. We learned that my
precious wife had a malignant tumor in her abdomen. We
consulted the finest doctors that we could find, and she took a

chemotherapy treatment. All of the ingenuity of modern medicine was thrown against the problem. But in the end it all seemed useless. In 1977, my wonderful wife Mary Wheeler Wigner passed away.

I found myself terribly lonely without her, eating in restaurants, silent and alone. I knew that there must be some advantages to being without a wife, but I could not think what they might be. I was without my closest confidante of so many years. After all the centuries, there is still a great taboo against speaking directly of death. I felt that taboo taking revenge on me.

At nearly the same time, Professor Donald Hamilton of the Princeton physics department also passed away. Dr. Hamilton had been a well-spoken man and a superb experimental atomic physicist. He had been chairman of the Princeton physics department until quite near the end.

I had known and admired him greatly, and also his wife, Eileen Hamilton, who had always been devoted to the physics department and to Princeton University. Over time, Mrs. Hamilton and I became much better friends, and soon I felt we should marry. Happily, she agreed. In 1979, I was married for the third and final time.

Eileen Hamilton Wigner is a lovely woman, a loving, trusting companion, a constant helper and friend. Many evenings, we play a kind of solitaire game for two. We lay out the cards: three, three, three, three; four, four, four. Very homelike. It is very nice to play on the same side of the table. I enjoy our marriage, and I believe that she does, too.

❧        ❧        ❧

Love does not weaken with age. But memory does. My loss of memory often seems to me a scandal and a very great weakness. Many of my remaining memories are trivial, while much that is important has flown from my mind.

Simple Hungarian poems and songs that I learned before 1910 still come to me unbidden. They are such darlings. But quite often now I recall a poem, but not its author; I begin thinking in Hungarian and cannot find my English again. I am amazed to find that sometimes even the name Adolf Hitler escapes me.

But if the failure of memory irks me, it also intrigues and pleases me. For the process of forgetting can also be a great and subtle pleasure. Few young people understand this. The pleasure of forgetting comes in ridding oneself of a piece of rudeness from years past.

Though most of my associates and acquaintances in life have been friendly, able, and cooperative, a few have been brooding, impolite, and not at all helpful. As a young man, my memory was sharp, and I could not help but recall specific instances of their unkindness, with shame and distaste. Now I do not.

When I returned from Louisiana to be a professor emeritus at Princeton, I had to decide what to study. I was, so to say, retired. A professor emeritus has very few official duties. But scholarship is not official at heart. True science cares nothing for the official arrangements of human institutions; it cares about such things as protons and neutrons. I still felt myself a scholar, with a duty to keep abreast of all theoretical physics, and indeed of all science.

I was still receiving 25 letters a day. Most of them asked bluntly for my money. But some of them made very complex requests, which I felt obliged to answer in some way. I also tried to correspond with many of my scientific colleagues. And I felt a duty to write to various politicians, telling them to defend our country. Without formal teaching duties and with enough money to support my family, I could easily have spent my retirement doing little more than reading and writing letters.

But I had broader interests. To think intellectually is a wonderfully human trait. My dog has no interest in the Associative Law of Multiplication. In my last years, I wanted to be fully human and to explore deeply the distinctly human quality of our life and thought.

I wanted time to savor the primary human realizations: that sexual intercourse creates children; that wild wheat can be domesticated; that food satisfies hunger; that a wheel transforms a cart, and language transforms human relations. But amidst this savoring, I felt a sadness at how little science I really knew. I knew a dog from a cat, but little more zoology than that. And like most of the best physicists, I knew little botany, medicine, astronomy, geology, metallurgy, or magic.

And I saw that this problem was widespread: Everywhere, mathematicians and physicists hardly knew chemistry; chemists did not know physics. Having been both a chemist and a physicist, I could say that.

All of my life I had wondered: How do we control our breathing? Perhaps a simple question to a physician, but I wanted for the first time to learn the answer and to somehow integrate it into my scientific thinking.

And what about psychology, that subject which had first enchanted me in Berlin 50 years before, by giving structure to the mind? Despite my scientific honors, I hardly knew psychology. And I was convinced that psychology is related to physics in ways that we scarcely understand.

Old men have a weakness for generality and a desire to see structures whole. That is why old scientists so often become philosophers. I had seen it happen to Einstein and Polanyi. By 1973, Heisenberg was largely a philosopher, and I was becoming one, too.

Modern philosophy dismayed me, clinging to lifeless categories like "Ethics" and "Epistemology," while declining to take on the meaning of human life, the motives and dilemmas of human society. The hectic modern world seemed to no longer allow philosophers leisure enough to paint a full picture in primary colors. I knew that I lacked the talent to change the whole direction of modern philosophy. So I decided to concentrate on physics.

I looked again at my life, and wondered: To what scientific problem should I devote my remaining years? I often thought of the colloquia at the University of Berlin in 1921, and of the great physicists in the front row, who sat with Einstein and wondered if man would ever solve the deepest questions posed by quantum theory. By 1973, we had known for many years that we could solve these questions. But by then we had new doubts. We doubted man could unite his knowledge of consciousness with that of physics.

Werner Heisenberg had told us 45 years before that the link between a man and his observation is a probability. That remarkable idea described the external world from an implicitly internal, psychological viewpoint. I felt that if scientists would explore this more deeply, we could surpass Heisenberg's uncertainty principle.

I wondered; Why are physics and the natural sciences kept separate? Why have we divided the world into physics and psychology, with physics describing external nature and psychology the mind's inner workings? Why not unite these two worlds in a science that addresses the meaning of emotion and memory?

Many physicists insist that nothing exists besides matter.

But our thoughts, our desires, and emotions—what are they then? If all that exists in my brain are a chain of complex chemical processes, why do I care what those processes are?

At the great colloquia in Berlin in 1921, I had expected science to soon begin unlocking the secrets of human society and human emotions. Fifty years later, it had vastly improved our material conditions, but scarcely examined human emotion.

Physics pretended to describe the workings of all the world: every property, every behavior. But the two most basic theories in physics, quantum mechanics and relativity theory, had never been truly united. I admired profoundly the practical success of quantum mechanics. But I felt that its reliance on inanimate objects was a major blemish.

And physics had not incorporated any of the miraculous world of biology. No one knew, for example, how far along the line from bacteria to humans the phenomena of pain and pleasure began to exist.

My chief scientific interest in the last 20 years has been to somehow extend theoretical physics into the realm of consciousness. Some scientists call consciousness a commonplace, and a subjective one at that. But consciousness is beautifully complex. It has never been properly described, certainly not by physics or mathematics. It is shrouded in mysteries. And what I know of philosophy and psychology suggests that those disciplines have never properly defined consciousness either.

Well, few scientific phenomena are well understood at first. Light was once called a stream of particles moving by mechanical laws. Electromagnetism was inscrutable without the concept of fields. Cosmology was mysterious before Einstein's general theory of relativity. Microscopic phenomena needed quantum mechanics to be understood. Even my precious field of symmetry was shaken in the late 1950s, and had to be broadly reordered.

To make quantum mechanics predict and explain consciousness and life, we will need new concepts that no longer rely chiefly on physics and chemistry. People tell me, "Eugene, consciousness is hardly physics." Well, perhaps so.

But consider the past. Classical physics said that matter could be divided infinitely without change in gravity, viscosity, or elasticity. When we found the atom, many scientists looked on it as an annoying complication.

When atomic theory arose to explain the atom, some physicists complained that this was "hardly physics." But it survived, was refined, and has by now beautifully enriched our picture of nature. Perhaps this familiar pattern will recur with consciousness.

I study these questions nearly every day, in my study at home, on my walks, in my workroom at Princeton University, sometimes even in my dreams. Physical ideas come to me rarely and at odd moments. But I continue to work.

I still like to study, to discover some foolish idea, even to write a paper if I come to it. I read mostly physics, but I still read a few novels and like to tap out Hungarian folk songs at the piano.

✿          ✿          ✿

One thing that still perturbs me is that we have never yet made contact with intelligent life on other planets. This would certainly be dangerous, but it should have occurred by now.

There are billions of stars and many of them have planets like ours. The universe should hold many other advanced life forms. They might well be dangerous, but how interesting they would be. So why have none of these creatures used their science to approach us?

I know two theories that might explain this. One is that advanced living creatures finally learn everything; when learning can no longer divert their quarrels, they turn their intelligence to making the weapons that finally destroy them and their whole world with them.

But man can die out without monstrous weapons. I believe that material wealth could become so nearly universal that advanced life forms would lose their sense of purpose, abandon reading, science, and most of the things that give life meaning. If they no longer understand life as something noble and worth striving for, they might stop making children.

Many scientists want to visit outer space, to seek out other creatures. I do so only in fantasy. I have too weak a stomach for space travel; it would surely make me seasick. I cannot feel that I belong in space.

The world beyond our outer atmosphere remains very foreign to me, while the affairs of this globe are very close and dear. Speculation about science and our earth's future is as tender to me as imagining the fate of a loved one.

❈   ❈   ❈

Physicists are not famous in America. Einstein was famous, but there has only been one Einstein. Most physicists are obscure. But among physicists there exists a smaller circle of fame. And there my article on, for instance, the phase-space description of quantum mechanics is well known.

The purpose for which I first created it is known; and in the many years since, it has been adapted to many fine purposes. To be the author of this article is to possess a certain kind of fame, and I must admit that I like it. I do not shun the admiration of physicists.

But I hate to consider myself an important figure. There is a certain kind of old-timer who enjoys playing the role of wise man. We have all seen this man. He courts professional attention, grows a white beard, and makes solemn pronouncements at regular intervals. The element of pretension in this role is repulsive to me. I cannot help feeling that living that way is somehow damaging to the soul.

My greatest success in life has been very personal: helping to produce two children. My lovely daughter, Martha, now has two darling daughters herself. She reminds me of my mother. There is no strong likeness between them, but no pronounced difference either. Martha has been well educated, but her main occupation now is to be a wife and mother, and to be my daughter. And she does that beautifully.

I have received many personal honors, not only the Nobel prize, but the Enrico Fermi award from the Atomic Energy Commission and the Banner Order of the Hungarian People's Republic. I have published over 500 articles. A few of my scientific ideas have been embraced as vital and will survive me.

Other physicists have developed these ideas, giving them life and color. For example, my work with isotopic spin and R-matrix theory has been remarkably satisfying. I have had a string of wonderful mentors, a few protégés, and many loyal friends. I am largely content. While a Russian of my generation can scarcely grasp the meaning of freedom, I do understand freedom, and am deeply gratified by it.

❋        ❋        ❋

Americans have often asked me, "Dr. Wigner, what do you think of our culture?" Well, I am poorly qualified to judge. After 60 years in the United States, I am still more Hungarian than American. I speak English with a thick accent. Much of the culture escapes me.

But the truth is that I find American culture a bit childish compared to German, Hungarian, French, or British culture, which look back so much farther. Americans do have Mark Twain and his *Huckleberry Finn,* but too few other great writers.

It is not true, as many Europeans say, that there are more fools in the United States than in Europe. Foolish people are evenly distributed. But conversation is taken less seriously here than in Europe, and very few children are taught to express themselves with much subtlety. And North Americans seem less curious than Europeans about the history, language, and culture of other nations.

Also, people settle more sparsely in America than in Europe and tend to think much less about caring for the land. That is a pity, because the American landscape is beautiful and deserves better care.

Americans seem to spend much of their lives watching television. I hardly watch; my wife does a bit more. On one subject, nearly all scientists agree, regardless of their politics: that the popular dramas on daytime television are abysmal. I do not agree. Television is very personal. Its programs are designed to describe popular desires, hopes, and fears. And they do this very well.

Americans love new inventions and often ask me to praise the modern computer. But compared to the wheel or the steam engine, I am afraid the modern "computer revolution" is a bit overdrawn. Modern computers are quite impressive. Computational speed can resolve great problems. And computers have certainly shaped the broader culture.

But in the study of science, I have never known a problem solved by a computer that could not have been solved without one. And that is the mark of a revolutionary advance: that it explains what has been inexplicable.

I feel that Americans would do better to teach every child the rudiments of physics and chemistry than to trust in the computer. It is sad how few people know the pleasure of practicing science. They say, "Oh, I have more practical interests than science," failing to see how broadly science touches their lives.

There are too many famous people in America, I think. They trip over each other, crowd and contradict one another. They know very well that they are famous. But few of them truly understand life deeply, as a great person should.

I find the American mind too often fixed on material things. Beauty, love, and attraction—these are life's greatest mysteries. To walk in a flowering garden, to bask in the sun, to embrace one's lover, to respect one's parents and delight in one's children—these are the truest joys in life.

Perhaps too few Americans realize this. But how do I know? I am loath to criticize too sharply a country that has adopted me and given me a safe and pleasant home for 60 years. One thing that Americans do understand deeply is freedom, and that is a very great blessing.

❄        ❄        ❄

On February 1, 1976, Werner Heisenberg died in Munich. After the war, Heisenberg had organized and directed the Max Planck Institute for Physics and Astrophysics at Göttingen. In 1958, he and his institute had moved to Munich. I was never an intimate of Heisenberg, but unlike some scientists, I did not resent the role he had played in Hitler's war. And I missed him.

Just three weeks after Heisenberg's death, my wonderful teacher and friend Michael Polanyi died. I have hardly mentioned Polanyi since Hitler drove him from Berlin to England in 1933. But in the 40 years after, he was never far from my thoughts.

Polanyi had taken up economics around 1939. By 1946, he had become largely a philosopher. Perhaps philosophy was Polanyi's true vocation. In 1958, he moved to Oxford. He felt a bit frustrated that his personal philosophy was not widely shared there.

But Polanyi's thought mixed science, aesthetics, and prophesy, and the distinctive, calm confidence with which he did this was slightly disquieting. Philosophy takes hold more slowly than pure science. I am not sure Polanyi knew that. He was a gentle man who felt sharply even mild criticism of his work.

Polanyi's last years were clouded with loss of memory, but he retained his rare sweetness of temper to the end. We often mailed each other personal manuscripts on scientific topics. Atrophy was a theme that concerned Polanyi deeply. Aware that skills are lost through disuse, he tried mightily to use all of his talents. He was always learning, incessantly.

Polanyi and I rarely saw each other, but I was often struck by the congruence of our deepest ideas and attitudes. As a young man, I was conscious of learning a fortune from Polanyi. I must have gleaned a great deal more subconsciously.

Even as an old man, when Polanyi had forgotten some of the riches that he had once known, he was still deft with praise. He used to respond to my manuscripts with lines like, "You will realize that your mathematical argument here is far above my head." His letters combined wisdom with great charm, which is not easy to do.

Polanyi's philosophical works had by then begun to be widely recognized. But it hurts me that Michael Polanyi is not more famous today, especially as a scientist. He certainly had a legion of admirers during his life. I was one of his pupils to the end.

And on October 20, 1984, my friend and brother-in-law Paul Dirac passed away. His departure from this world made a widow of my sister Manci and brought me more sorrow than I can adequately describe.

At least I still have Edward Teller. Of course, Teller is still criticized. His most recent "mischief" is in promoting the Star Wars program. Teller and I no longer speak more than a few times a year. He lives in California, I travel less than I once did, and Edward is always terribly busy. But he is Hungarian, and old Hungarian friends never completely lose touch. Our occasional talks mean a very great deal to me.

I am very lucky, and I like to be. I do not want to die yet. Remarkable things are happening in the world. The Russian leaders no longer dream the old communist dream of world conquest. They see that this dream is not shared by their own people. They may hate the United States military, but they do not expect it to attack Russia. This is real progress, and it encourages me.

The Russians see that the Marxist idea of the world as one communist society is wrong. They seem to see that countries live different lives, and that many cultures should compete for man's allegiance.

Certainly they permit greater emigration. Only a few of the ancient Roman emperors yearned to subdue all the peoples in their empire. The later Romans saw that they could no longer rule a huge empire from Rome, and they divided it up. Today it seems that the Russians will do the same.

Apparently, Hungary is now largely free. But I have no desire to move back there. It is not at all the country that I left 60 years ago. My family is in the United States, and most of my friends and closer colleagues, too. Princeton University is rooted in New Jersey, and also my better half. I doubt that either one of them would consent to move to Budapest.

❉      ❉      ❉

Life can be bitter at the end, but not yet for me. Sometimes I feel dizzy when I rise. And the state of New Jersey no longer lets me drive my car. But I can still get around fairly well. I still take a five-minute walk every day and dream of new areas that physics could explore. I am still six years older than Edward Teller and nearly as vigorous.

Many old people quarrel with their doctors and complain of medical ignorance and arrogance. But I have found most doctors to be humble and possessed of great wisdom. Few people today realize how far we have come from the days of my grandfather Einhorn, when so many people died prematurely.

Death can come sooner and it can come later. All that you can say is: "I am fine, as far as I know." People say that death is too terrible to discuss. But why? Life is wonderful, but we all will die eventually. Two of my wives are dead. Many of my Hungarian countrymen were killed by Hitler. All of my principal mentors and most of my closest friends have passed away.

That I will die hardly bothers me. What you cannot resist slips from your consciousness, like an idea you cannot grasp. The only trouble with death is the pain it brings to friends and family. But we are all guests in this world, and our culture commits a crime when it persuades us to think otherwise.

People say that Heaven awaits some of us. Now, *that* is

something incomprehensible. No one can speak factually of Heaven. As a scientist, I must say that we have no heavenly data. So I am afraid that after death, we merely cease to exist.

But I am unsure. It is touching how badly people want to know their soul's ultimate fate. I do not mind speculating and waiting. Someone once asked me what I would say to Leo Szilard if he walked in my front door tomorrow. Only this: "How did you get back from the wonderful world in which you have been?"

I scarcely feel that I have understood life. But then human beings are hugely innocent. We may detect many of the physical laws of our world and observe many of the desires and traits of its inhabitants. But we never see the thing whole. And we never see very far ahead.

The full meaning of life, the collective meaning of all human desires, is fundamentally a mystery beyond our grasp. As a young man, I chafed at this state of affairs. But by now I have made peace with it. I even feel a certain honor to be associated with such a mystery.

I have tried to be happy in this life; to keep male and female friends; to help my family. I could not do much better with a second chance. Surely, I could be a better son, husband, and father. I could support my teachers better. I have tried many things in life, a few of them immoral. But if I have done anything truly wicked, it no longer disturbs my mind.

" 'The future is uncertain' says the optimist." That dryly skeptical proverb is a favorite of mine. But we should not paint the future too darkly. And I hope very strongly that everything will turn out all right.

And now I think you know more about me than is reasonable.

# ℬ𝒾𝒷𝓁𝒾𝑜𝑔𝓇𝒶𝓅𝒽𝓎

## ℬ𝑜𝑜𝓀𝓈

Bainbridge, Kenneth. *Trinity.* Los Alamos, N. M.: Los Alamos Scientific Laboratory, 1945 (1976).

Blumberg, Stanley A., and Gwinn Owens. *The Life and Times of Edward Teller.* New York: G. P. Putnam's Sons, 1976.

Born, Max. *My Life: Recollections of a Nobel Laureate.* New York: Charles Scribner's Sons, 1975.

Bundy, McGeorge. *Danger and Survival: Choices about the Bomb in the First Fifty Years.* New York: Random House, 1988.

Clark, Ronald. *Einstein: The Life and Times.* New York: Avon, 1972.

Compton, Arthur Holly. *Atomic Quest: A Personal Narrative.* New York: Oxford University Press, 1956.

Conant, James Bryant. *A History of the Development of an Atomic Bomb.* Unpublished manuscript, Bush-Conant File, Folder 5, National Archives.

Davis, Nuel Pharr. *Lawrence and Oppenheimer.* New York: Simon and Schuster, 1968.

Fermi, Laura. *Atoms in the Family.* Chicago: University of Chicago Press, 1954.

Ford, Daniel. *The Cult of the Atom: The Secret Papers of the Atomic Energy Commission.* New York: Simon & Schuster, 1982.

French, A. P., ed. *Einstein: A Centenary Volume.* Portsmouth, N. H.: Heinemann Publishers, 1979.

Heims, Steve J. *John von Neumann and Norbert Wiener: From Mathematics to the Technologies of Life and Death.* Cambridge, Mass.: MIT Press, 1980.

Hewlett, Richard G., and Oscar Anderson, Jr. *The New World, 1939/1946.* University Park: Pennsylvania State University Press, 1962.

Hewlett, Richard G., and Francis Duncan. *Atomic Shield, 1947–1952.* University Park: Pennsylvania State University Press, 1969.

Jungk, Robert. *Brighter Than a Thousand Suns.* San Diego: Harcourt Brace, 1958.

Kevles, Daniel. *The Physicists.* New York: Alfred A. Knopf, 1978.

Kurzman, Dan. *Day of the Bomb: Countdown to Hiroshima.* New York: McGraw-Hill Book Company, 1986.

Libby, Leona Marshall. *The Uranium People.* New York: Crane Russak & Company, Charles Scribner's Sons, 1979.

Lukacs, John. *Budapest 1900: A Historical Study of a City and Its Culture.* New York: Grove Weidenfeld, 1988.

Mark, Hans, and Sidney Fernbach, eds. *Properties of Matter under Unusual Conditions.* "An Appreciation on the 60th Birthday of Edward Teller," by Eugene P. Wigner. New York: Interscience, 1969.

Meystre, P., and M. O. Scully, eds. *Quantum Optics, Experimental Gravity and Measurement Problem.* "The Glorious Days of Physics," by Eugene P. Wigner. New York: Plenum Press, 1983.

Moss, Norman. *Men Who Play God.* New York: Harper & Row, 1969.

Muses, Charles, and Arthur M. Young, eds. *Consciousness and Reality: The Human Pivot Point.* New York: Avon, 1974.

Nichols, K. D. *The Road to Trinity.* New York: William Morrow & Company, 1987.

Rhodes, Richard. *The Making of the Atomic Bomb.* New York: Simon and Schuster, 1986.

Sayen, Jamie. *Einstein in America.* New York: Crown Publishers, Inc., 1985.

Sherwin, Martin. *A World Destroyed.* New York: Alfred A. Knopf, 1975.

Shirer, William. *Berlin Diary.* New York: Alfred A. Knopf, 1941.

Smith, Alice Kimball, and Charles Weiner, *Robert Oppenheimer: Letters and Recollections.* Cambridge, Mass.: Harvard University Press, 1980.

Strauss, Lewis. *Men and Decisions.* Garden City, N.Y.: Doubleday & Company, 1962.

Stuewer, Roger H., ed. *Nuclear Physics in Retrospect.* "The Neutron: The Impact of Its Discovery and Its Uses." Minneapolis: University of Minnesota Press, 1979.

Sylves, Richard T. *The Nuclear Oracles: A Political History of the General Advisory Committee of the Atomic Energy Commission.* Ames: Iowa State University Press, 1987.

Teller, Edward. *The Legacy of Hiroshima.* New York: Doubleday, 1962.

Wigner, Eugene P., ed. *Physical Science and Human Values.* Princeton, N.J.: Princeton University Press, 1947.

Wigner, Eugene P. *Symmetries and Reflections,* reprint ed. Woodbridge, Conn.: Ox Bow Press, 1979.

Woolf, Harry, ed. *Some Strangeness in the Proportion.* "Thirty Years of Knowing Einstein," by Eugene P. Wigner. Redding, Mass.: Addison Wesley, 1980.

## *Articles*

Goldschmidt, Bertrand, moderator. "The Discovery and Contribution of Nuclear Fission." International Conference on Nuclear Fission: Fifty Years of Progress in Energy Security. *Transactions of the American Nuclear Society* 59, 1989.

Goldstine, Herman H., and Eugene P. Wigner. "The Scientific Work of John von Neumann." *Science* 125:683, 1957.

Ladenburg, Rudolf, and Eugene P. Wigner. "Award of the Nobel Prizes in Physics to Professors Heisenberg, Schrödinger and Dirac." *Scientific Monthly* 38:87, 1934.

Weinberg, Alvin M., and Eugene P. Wigner. "Longer Range View of Nuclear Energy." *Bulletin of the Atomic Scientists* 16:10, 490, 1960.

Wigner, Eugene P. "On the Quantum Correction For Thermodynamic Equilibrium." *Physical Review* 40:749, 1932.

Wigner, Eugene P. "On Unitary Representations of the Inhomogeneous Lorentz Group." *Annals of Mathematics* 40:149, 1939.

Wigner, Eugene P. "Are We Making the Transition Wisely?" *Saturday Review* 28, November 17, 1945.

Wigner, Eugene P. "Theoretical Physics in the Metallurgical Laboratory of Chicago." *Journal of Applied Physics* 17:857, November, 1946.

Wigner, Eugene P. "Atomic Energy." *Science* 108:517, 1948.

Wigner, Eugene P. "The Limits of Science." *Proceedings of the American Philosophical Society* 94:422, 1950.

Wigner, Eugene P. "Enrico Fermi." *Year Book of the American Philosophical Society* 435, 1955.

Wigner, Eugene P. "The New Editor of *Reviews of Modern Physics,* E. U. Condon." *Physics Today* 30, November, 1956.

Wigner, Eugene P. "John von Neumann (1903–1957)." *Year Book of the American Philosophical Society* 1957, 149.

Wigner, Eugene P. "Conference on Invariance." *Physics Today* 78, March, 1960.

Wigner, Eugene P. "The Unreasonable Effectiveness of Mathematics in the Natural Sciences." *Communications in Pure and Applied Mathematics* 13:1, 1960.

Wigner, Eugene P. "Recall the Ends—While Pondering the Means." *Bulletin of the Atomic Scientists* 17:3, March, 1961.

Wigner, Eugene P. "Union of German Physical Societies." *Physics Today* 15:98, 1962.

Wigner, Eugene P. "Review, *American Scientists and Nuclear Weapons Policy,* by Robert Gilpin." *Bulletin of the Atomic Scientists* 29, October, 1962.

Wigner, Eugene P. "Review, *The Inspiration of Science,* by Sir George Thompson." *Bulletin of the Atomic Scientists* 27, November, 1962.

Wigner, Eugene P. "Fermi Award: A.E.C. Honors Teller for Contributions to Nuclear Science." *Science* 138:1087, 1962.

Wigner, Eugene P. "The Problem of Measurement." *American Journal of Physics* 31:1, January, 1963.

Wigner, Eugene P. "Twentieth Birthday of the Atomic Age." *New York Times Magazine* 34, December 2, 1963.

Wigner, Eugene P. "Two Kinds of Reality." *The Monist* 48:248, 1964.

Wigner, Eugene P. "Why Civil Defense?" *Technology Review* 66:8, June, 1964.

Wigner, Eugene P. "Events, Laws of Nature and Invariance Principles." (Nobel Address, The Nobel Foundation, Stockholm). *Science* 145:120, 1964.

Wigner, Eugene P. "Letter: On Article by Wiesner and York." *Scientific American* 211:6, December, 1964.

Wigner, Eugene P. "Does Quantum Mechanics Exclude Life?" *Nature* 205:1306, 1965.

Wigner, Eugene P. "Civil Defense: Wigner on Project Harbor." *Bulletin of the Atomic Scientists* 22:2, February, 1966.

Wigner, Eugene P. "Book Review: *Of Molecules and Men,* by Francis Crick." *Science* 156:798, May 12, 1967.

Wigner, Eugene P. "Letter: Djilas Forgiven." *Princeton Alumni Weekly* 69:3, February 4, 1969.

Wigner, Eugene P. "Are We Machines?" *Proceedings of the American Philosophical Society* 113:2, April 17, 1969.

Wigner, Eugene P. "Leo Szilard, 1898–1964." *Biographical Memoirs* 40, Columbia University Press, 1969.

Wigner, Eugene P. "The Myth of Assured Destruction." *Survive* 3:4, July–August, 1970. (Reprinted in *Congressional Record,* E920, February 19, 1971.)

Wigner, Eugene P. "On Some of Physics' Problems." *Main Currents in Modern Thought* 28:75, 1972.

Wigner, Eugene P. "Letter: Falk Debated." *Princeton Alumni Weekly* 7, February 27, 1973.

Wigner, Eugene P. "Letter: Wigner Replies." *Physics Today* 13, June, 1974.

Wigner, Eugene P. "Obituary: Werner K. Heisenberg." *Physics Today* 86, April, 1976.

Wigner, Eugene P., John Jewkes, and Rom Harré. "Obituary: Michael Polanyi." *Nature* 261:83, May 6, 1976.

Wigner, Eugene P. "Book Review: *The Advisors: Oppenheimer, Teller and the Superbomb,* by Herbert York." *American Scientist* 64:561, September–October, 1976.

Wigner, Eugene P., and R. A. Hodgkin. "Obituary: Michael Polanyi." *Biographical Memoirs of the Royal Society* 23:413, 1977.

Wigner, Eugene P. "Book Review: *Albert Einstein, The Human Side,* by Helen Dukas and Banesh Hoffman." *Nature* 282:179, 1979.

Wigner, Eugene P. "Einstein's Ideals." *Nature* 282:179, 1979.

Wigner, Eugene P. "The Limits of Science." *The World and I* 304, February, 1986.

Wigner, Eugene P. "Letter: More on Civil Defense." *Physics Today* December, 1986.

### *Public Addresses*

"Remarks at the Designation of the X-10 Graphite Reactor as a National Historic Landmark." Address by Eugene P. Wigner, Oak Ridge National Laboratory, September 13, 1966.

"The Miracle of Science." Address by Eugene P. Wigner, University of Lowell, April 16, 1985.

### *Newspaper Articles*

Weinberg, Alvin, and Eugene P. Wigner. "New Light on the Einstein Letter." *Oak Ridger,* March 11, 1986.

Wigner, Eugene P. "Book Review, *Einstein: The Life and Times,* by Ronald Clark." *The Philadelphia Inquirer* 23, September 22, 1971.

Wigner, Eugene P. "Why I Plan To Vote Republican." *Princeton Town Topics,* 27, October 26, 1972.

Wigner, Eugene P. "Death or Surrender? I Prefer Freedom." *New York Tribune,* April 27, 1983.

"Science Should Study Emotions Says Wigner." *Oak Ridger* 1, May 14, 1975.

"Nobel Laureate in Physics Looks toward Future." *Baton Rouge Morning Advocate,* January 1, 1986.

"A-Bomb Developer Backs Star Wars Defense." *Knoxville News-Sentinel,* February 16, 1986.

### *Interviews*

Interview with Eugene P. Wigner, by I. Kardos. *Scientists Face to Face.* Corvina Books, Budapest, 1978.

Interview with Eugene P. Wigner, by Thomas Kuhn, 1963, Archive for the History of Quantum Physics, copy in The American Philosophical Society Library, Philadelphia.

Interview with Eugene P. Wigner, by Charles Weiner, 1966, Niels Bohr Library, American Institute of Physics, New York.

Interview with Eugene P. Wigner. *New Engineer Magazine* 1:2, November, 1971.

Interview with Eugene P. Wigner, "Ethics in Relationship between Science and Society." *Impact of Science On Society* 22:283, 1972.

Interview with Eugene P. Wigner. *New Hungarian Quarterly* 14:51, Autumn, 1973.

Interview with Eugene P. Wigner, by J. Walsh. *Science Magazine* 181:527, August 10, 1973.

*Papers*

Polanyi, Michael, and Eugene P. Wigner, correspondence, Michael Polanyi Papers, Department of Special Collections, University of Chicago Library.

Szilard, Leo, and Eugene P. Wigner, correspondence, Leo Szilard Papers (MSS 32), University Library, Mandeville Department of Special Collections, University of California, San Diego.

*Index*